Learning Math With Bible Heroes
Grade 3

This book has been designed to help children learn important math skills and some fascinating aspects of God's teachings at the same time. It addresses fundamental third grade math skills (memorization of facts, math procedures, etc.). It is equally important that children at this level understand the basic concepts involved in math.

Learning Math With Bible Heroes provides children with practice in concept development (understanding the process) and provides opportunities for skill practice as well. All activities are presented in an interesting format that is child-friendly.

As you might expect from looking at the title of this book, each activity presented features some aspect of God's teachings. Use these lessons as building blocks for further discussions. This book is a wonderful way to help children learn math concepts and skills using a religious format. It will also help children see math as an important part of our everyday lives.

How to Use This Book

- Provide a quiet, comfortable place in which you can work with the child.
- Create a warm, accepting atmosphere so that the child will feel successful.
- Make sure that the child understands the directions before beginning an activity.
- Offer praise and encouragement to the child throughout each activity.
- Check the answers with the child as soon as an activity has been completed. (Answers are provided on the pull-out pages in the center of the book.)
- Math topics included in this book are as follows: adding, subtracting, place value, problem solving, rounding, measuring, geometry, graphing, multiplication, division, time, fractions, mixed numbers, decimals, money, and many more.
- Cut out the manipulatives on pages 50-52. Have the child color them. Store the pieces in a sealable plastic bag.
- Reward the child for his or her hard work with the award on page D in the center of the book. This is a great way to boost self-esteem!

Addition/subtraction

Important People

The Bible contains lots of people with whom you are probably familiar. Below are the names of some important people mentioned in the Bible. Add or subtract the total number of letters in each name to get each answer.

1. Samson + Jonah = 6 + 5 = 11

2. Joseph + David + Goliath = _____

3. Moses + King Solomon = _____

4. Daniel – Noah = _____

5. Abraham – Joshua = _____

6. Nebuchadnezzar – Jacob = _____

7. Benjamin + Ruth = _____

8. Gideon + Sarah = _____

9. Cain + Abel + Lot = _____

10. Rebekah – Esau = _____

11. Isaac + Rachel = _____

12. Matthew – Judah = _____

13. Zechariah – Levi = _____

14. Caleb – Naomi = _____

15. Hannah + Samuel = _____

Addition/subtraction

Lions' Den

Daniel was thrown into the lions' den. Help him find his way out by following the path that equals 27. Write your answers as you move through the paths. Color Daniel's path green.

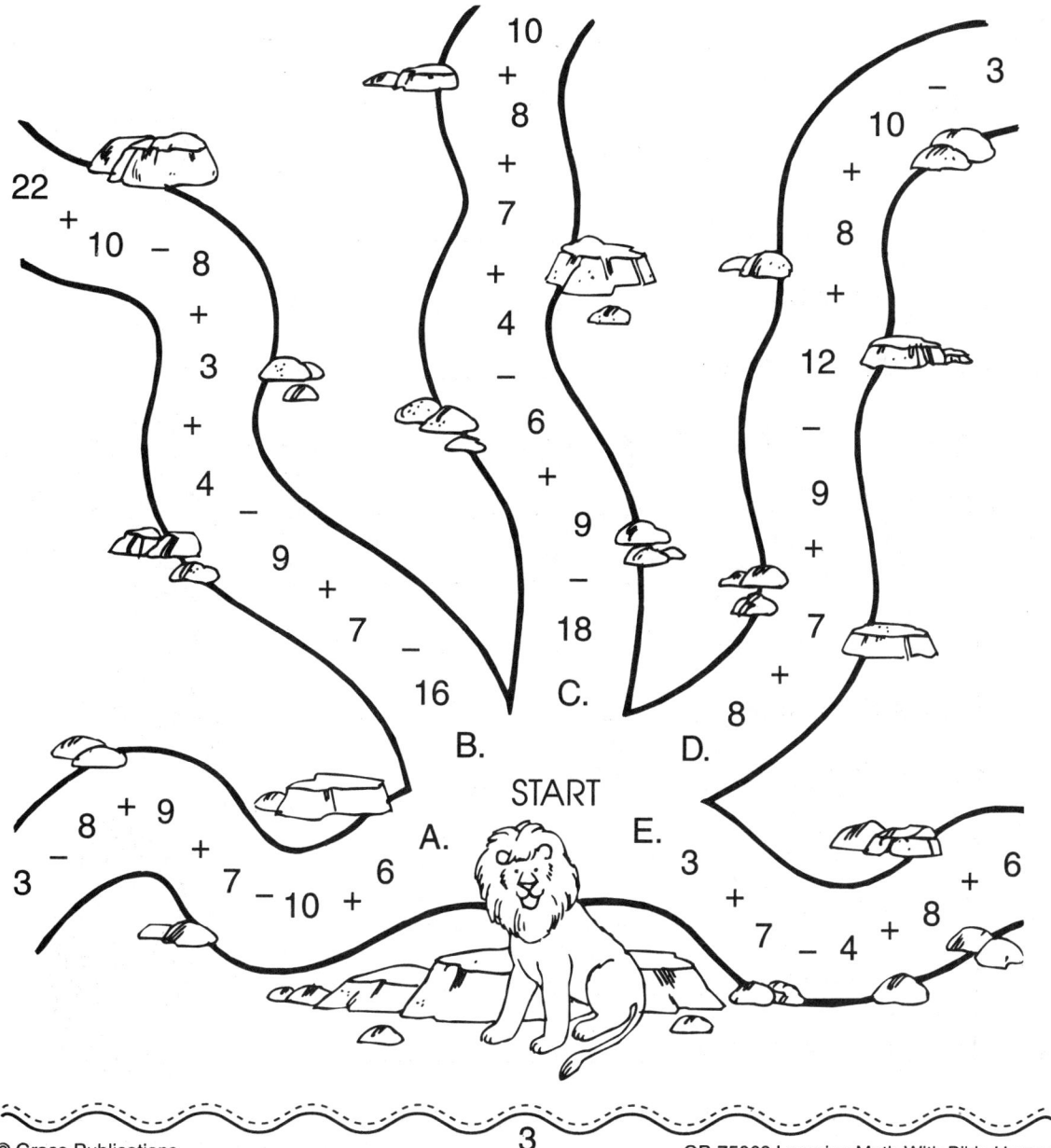

3 addends

Fishing for Addends

Long ago, many of God's people were fishermen. Add the numbers on each worm. Draw a line to the fish containing the same sum.

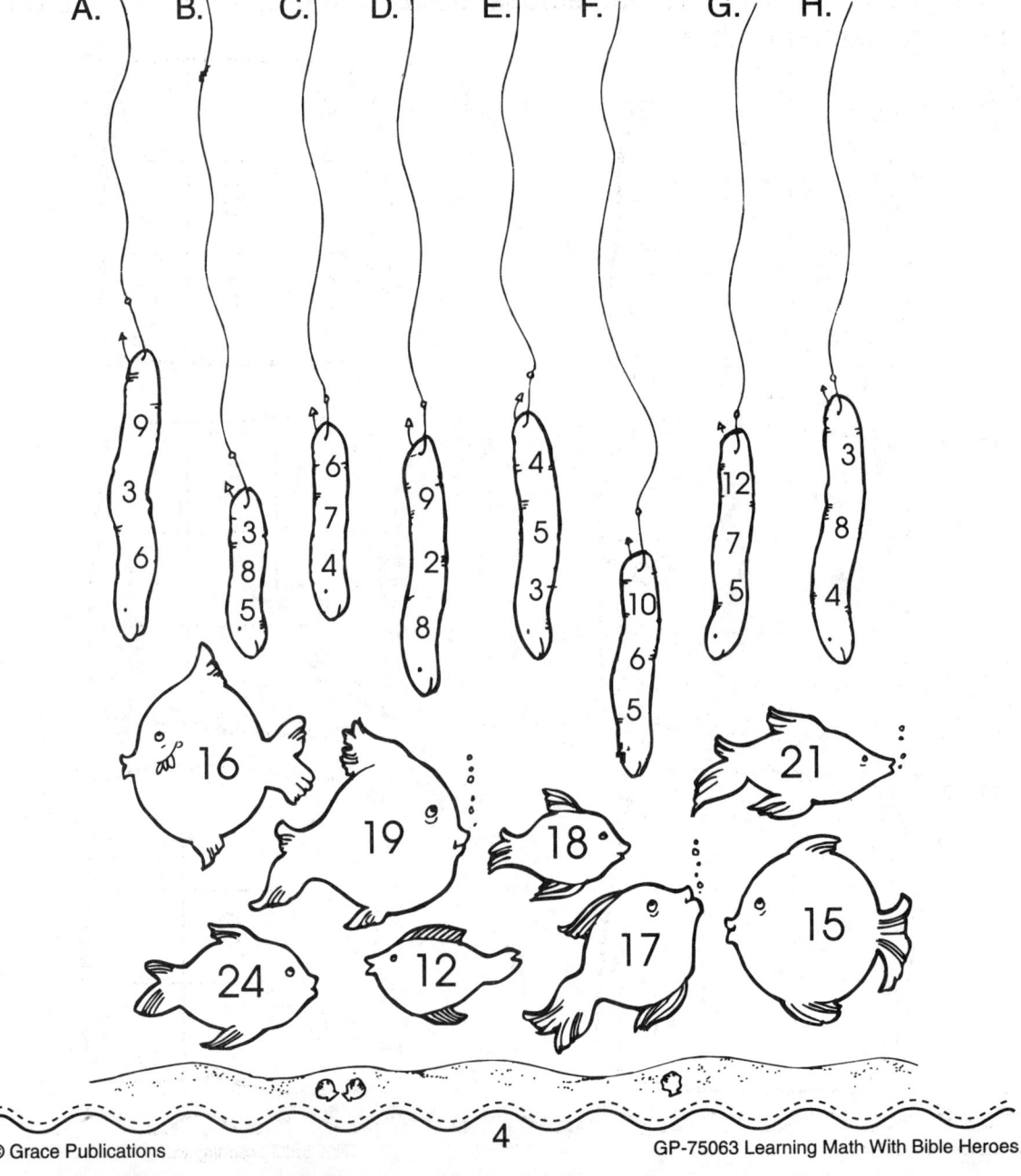

Samson's Squares

Addition/subtraction

Samson was a powerful young man who killed a lion with his bare hands. Show your addition power by filling in the missing numbers. All the numbers, when added across, down, and diagonally, should add up to the same number.

A. 24

11	4	9
6	8	10
7	12	5

B. 15

4		
	5	
	2	7

C. 30

9		
		12
13		

D. 27

	5	
		11
	8	

E. 21

4		
	7	
		10

F. 27

6		
	9	
	8	

Problem solving

Animals for the Ark

God told Noah to gather two of every kind of animal to put on the ark. Below are some of the animals Noah gathered. They are very hungry. Fill their tummies with numbers. The sum of each row must add up to the number in the animal's mouth.

Addition/subtraction

Mazes of Moses

Moses led the people of Israel through the wilderness.

Find your way through the number mazes below. Start with one number and add or subtract to get from one box to another. Each number will be used only once, but your paths may cross. Start in the box containing the dot. Write your maze of problems on the lines provided.

Example:

22	8	14
9	10	4
13	3	7•

7 + 3 = 10 + 4 =
14 + 8 = 22 − 9 = 13

A.

15•	6	21
4	27	7
19	8	28

B.

13	5	18
21	5	3
6	15	8•

C.

10	4	14
3	15	7
7•	8	7

D.

13	6	7
3	20	4
10	10	11•

E.

18	9	9
9	16	7
7	7	14•

© Grace Publications

7

GP-75063 Learning Math With Bible Heroes

Place value

Numbers

In the Book of Numbers, Moses lists all the families who left Egypt. Moses also wrote God's instructions for the Promised Land. Complete the puzzle below by writing the correct number for each group of number words.

ACROSS

1. four hundred seventy-three thousand, two hundred twenty-six
3. nine hundred forty-six
4. seven hundred thirty-three
5. four hundreds, eight tens, three ones
6. ninety-thousand, seven
7. seven thousand, eight hundred twenty
8. eight hundred nineteen thousand, six hundred thirty-four
10. six tens, four ones
11. three hundred twenty-two thousand, four hundred nine

DOWN

2. two ten-thousands, nine thousands, eight hundreds, eight ones
4. seven thousands, six hundreds, eight ones
5. four hundred seven thousand, four hundred thirty
6. ninety thousand, six hundred eighty-two
9. one hundred, nine tens, three ones

Place value

Ezra, the Leader

Ezra gathered a group of people still living in Babylon and led them back to Jerusalem. He cleaned the temple, read from the law of Moses, and taught priests how to make proper offerings. Help Ezra lead the people to Jerusalem by following the clues below.

Write six different numbers by changing the place value of 4, 8, and 7.

1. _____ 2. _____ 3. _____ 4. _____ 5. _____ 6. _____

Now, write the numbers in the blanks along the path. Arrange them from the least value to the greatest value. Start at Babylon and end at Jerusalem.

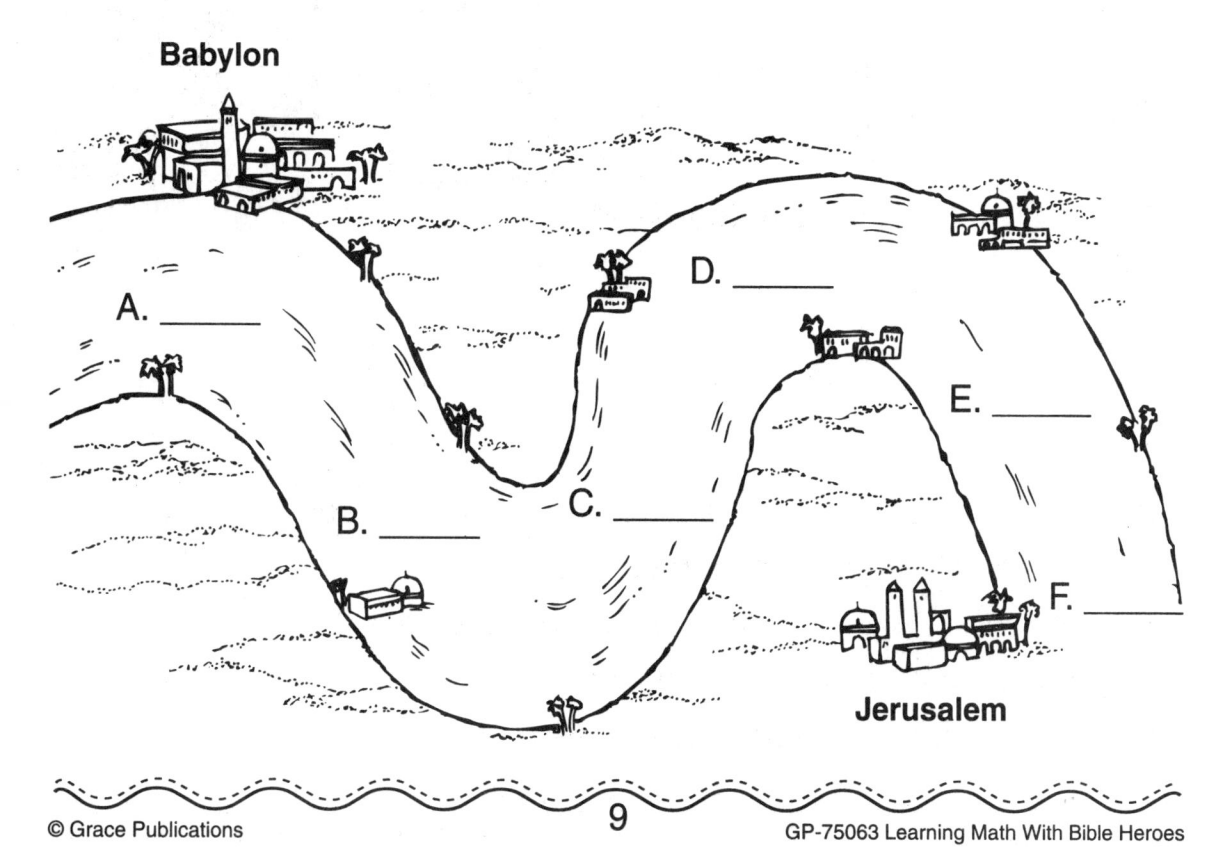

Babylon

A. _____

B. _____

C. _____

D. _____

E. _____

F. _____

Jerusalem

Rounding

Great Faith

One of the heroes in the Bible is very well known for his great faith in God. He followed the Lord's direction and moved from Haran to Canaan. He was the father of the ancient Hebrew nation. To find out who he was, solve the problems below. Then round each answer to the nearest ten. Write the matching letter in the blank above the answer to the problem at the bottom of the page. (Some letters will not be used.)

P.
   630
  −418

H.
   238
  +147

A.
   522
  −324

T.
   268
  +193

M.
   367
  +402

R.
   223
  +144

C.
   522
  +297

A.
   178
  − 32

A.
   310
  −176

B.
   703
  −462

___ ___ ___ ___ ___ ___ ___
150 240 370 130 390 200 770

10

© Grace Publications   GP-75063 Learning Math With Bible Heroes

Greater than/less than

Promises

God made promises with his people. These promises, or special agreements, had a special name. To find out what they are called, add or subtract each problem below. Then use > or < to show the greater number. Unscramble the letters under the greater numbers to fill in the answers in the blanks at the bottom of the page.

1. 8012   7691
   − 934   − 543
     E   N

2. 6573   5821
   − 842   − 907
     N   B

3. 4281   5430
   + 987   + 912
     R   O

4. 3047   5086
   − 951   − 843
     L   T

5. 1023   2765
   + 581   + 742
     D   C

6. 3807   5086
   + 938   + 793
     P   E

7. 942   731
   − 786   − 347
     B   S

8. 2376   837
   + 4298 + 426
     A   W

9. 312   846
   − 173   − 683
     I   V

___ ___ ___ ___ ___ ___ ___ ___

Time

Noah's Day

Noah obeyed God by building an ark and gathering animals to put in it. He had many animals to take care of. They all had to be fed at different times. Help Noah set the clocks so he can feed the animals.

1. 1:21   2. 3:47   3. 6:05   4. 4:35

5. three thirteen   6. seven twenty-four   7. two thirty-two   8. twelve o'clock

9. 2:06   10. half past six   11. 5:52   12. quarter past nine

© Grace Publications   GP-75063 Learning Math With Bible Heroes

Time

Time for Noah

Noah has lots to do to take care of the animals. Help him figure out a schedule so he can get everything done. Write the time shown on each clock two ways.

1. Heat milk.

_____
_____

2. Walk the lions.

_____
_____

3. Bathe the elephants.

_____
_____

4. Feed the parrots.

_____
_____

5. Clean tiger cages.

_____
_____

6. Water the plants.

_____
_____

7. Spread hay.

_____
_____

8. Scatter birdseed.

_____
_____

9. Light the lanterns.

_____
_____

© Grace Publications   GP-75063 Learning Math With Bible Heroes

Feeding the Animals

Money

Noah needs to buy food for the animals in the ark. Help him out by answering the questions below. Use the bills and coins from pages 50-52 to help you count the money.

1. What three things can Noah buy with 1 dollar, 3 quarters, 1 dime, and 3 pennies?

   _____

2. What two things can Noah buy with 2 quarters, 1 dime, 3 nickels, and 1 penny?

   _____

3. What four things can Noah buy with 2 dollars, 1 quarter, 4 dimes, 3 nickels, and 4 pennies?

   _____

4. What two fruits can Noah buy with 2 quarters, 3 dimes, 2 nickels, and 3 pennies?

   _____

5. What two things can Noah buy with 2 dollars, 5 quarters, 8 dimes, and 1 penny?

   _____

6. What four things can Noah buy with 2 dollars, 4 quarters, 4 dimes, 4 nickels, and 3 pennies?

   _____

Shopping With Noah

Money

Use the items on page 14 to help Noah decide if he has enough money to buy food for his animals.

1. Noah has 4 dollars, 3 quarters, 2 dimes, and 1 nickel. Noah has _____. Can he buy grain and seed? _____
2. Noah has 8 quarters, 4 dimes, 2 nickels, and 3 pennies. Noah has _____. Can he buy apples, grapes, and seed? _____
3. Noah has 5 quarters, 5 dimes, 3 nickels, and 2 pennies. Noah has _____. Can he buy carrots, fish, and raisins? _____
4. Noah has 3 quarters, 6 dimes, 1 nickel, and 4 pennies. Noah has _____. Can he buy corn, grapes, fish, and carrots? _____
5. Noah has 3 dollars, 4 quarters, 3 dimes, and 1 penny. Noah has _____. Can he buy carrots, apples, and grain? _____
6. Noah has 2 dollars, 3 quarters, 3 dimes, and 3 pennies. Noah has _____. Can he buy seed, corn, grapes, and fish? _____
7. Noah has 2 dollars, 5 quarters, 6 dimes, 1 nickel, and 2 pennies. Noah has _____. Can he buy grain and fish? _____
8. Noah has 1 dollar, 2 quarters, 4 dimes, 1 nickel, and 3 pennies. Noah has _____. Can he buy seed, raisins, and apples? _____
9. Noah has 3 quarters, 2 dimes, and 1 penny. Noah has _____. Can he buy carrots and corn? _____
10. Noah has 3 quarters, 2 dimes, 1 nickel, and 2 pennies. Noah has _____. Can he buy raisins, apples, and corn? _____

Money Mystery

Noah needs to pay $1.00 for some wood to build the ark. Noah only knows one way to make $1.00. Help Noah find more ways to make $1.00. Decide which coins to use. Use the coins from pages 50-52 to help you. Show each way of making $1.00 in the chart below.

Quarters	Dimes	Nickels	Pennies

Adding Giant Numbers

Addition

Goliath was a Philistine giant who was over nine feet tall. A boy named David was determined to slay Goliath. Help David by showing your addition strength. Using the code, add the numbers indicated according to their location in the puzzle.

87	785	867
493	234	48
645	376	549

1. 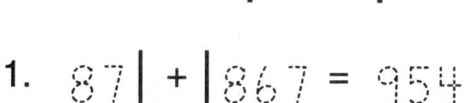 87 + 867 = 954
2. ⌐ + ⌐ =
3. ⌐ + ⌐ =
4. ⌐ + ⌐ =
5. ⌐ + ⌐ =
6. ⌐ + ⌐ =
7. ⌐ + ⌐ =

8. ⌐ + ⌐ =
9. ⌐ + ⌐ =
10. ⌐ + ⌐ =
11. ⌐ + ⌐ =
12. ⌐ + ⌐ =
13. ⌐ + ⌐ =
14. ⌐ + ⌐ =

Subtraction

David's Faith

David's faith in God helped him slay Goliath. Solve the problems below. Then use the code to find out what David used to kill Goliath. (Some letters will not be used.)

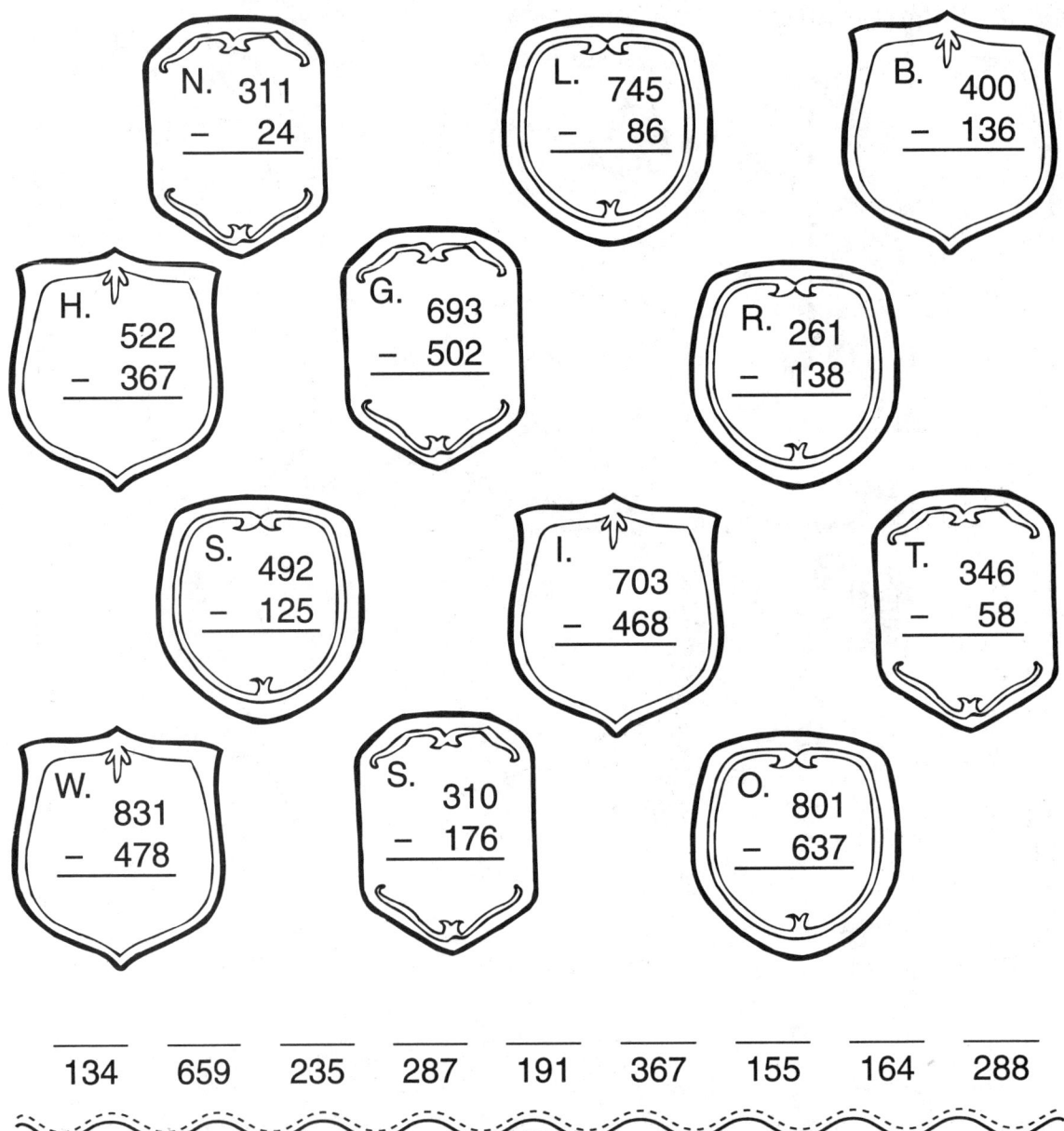

| 134 | 659 | 235 | 287 | 191 | 367 | 155 | 164 | 288 |

Subtraction

Goliath's Fall

David trusted his faith in God when he battled Goliath. Armed with only five smooth stones, he was able to make Goliath fall. Solve each problem below. Then use the code to find out where David's rock hit Goliath. (Some letters won't be used.)

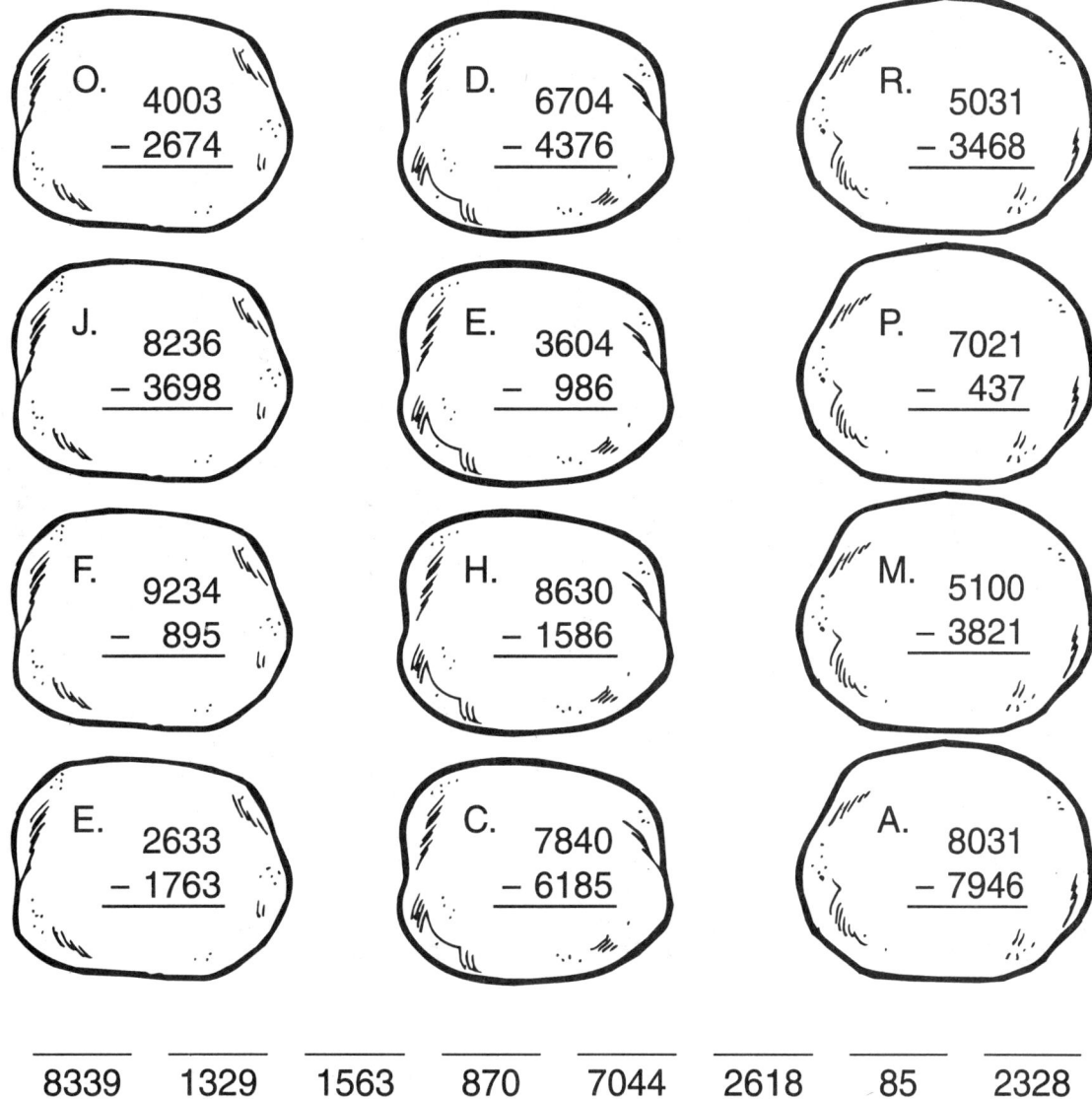

O. 4003 − 2674

D. 6704 − 4376

R. 5031 − 3468

J. 8236 − 3698

E. 3604 − 986

P. 7021 − 437

F. 9234 − 895

H. 8630 − 1586

M. 5100 − 3821

E. 2633 − 1763

C. 7840 − 6185

A. 8031 − 7946

____ ____ ____ ____ ____ ____ ____ ____
8339 1329 1563 870   7044 2618   85   2328

Measurement

Measure Me!

God made everyone special. You are unique! Find out about yourself by completing the activities below. You will need a measuring tape or a string and a ruler, a scale, and a clock to complete this page.

I am _____ inches tall.

I am taller than a _____.

I am shorter than a _____.

Ten of me would be _____ feet.

That's as tall as _____.

I weigh _____ pounds.

Twenty of me would weigh _____.

That's as heavy as _____.

I can hold my breath for _____ seconds.

I can jump _____ inches or _____ feet _____ inches high.

My arm is _____ inches long.

My leg is _____ inches long.

My neck is _____ inches around.

My wrist is _____ inches around.

My ankle is _____ inches around.

My waist is _____ inches around.

My nose is _____ inches long.

My ear is _____ inches long.

My head is _____ inches around.

Perimeter

Babel's Bricks

As the kingdom of Babel prospered, the people learned how to make bricks. They decided to build a tower to heaven. Find the perimeter of each brick below.

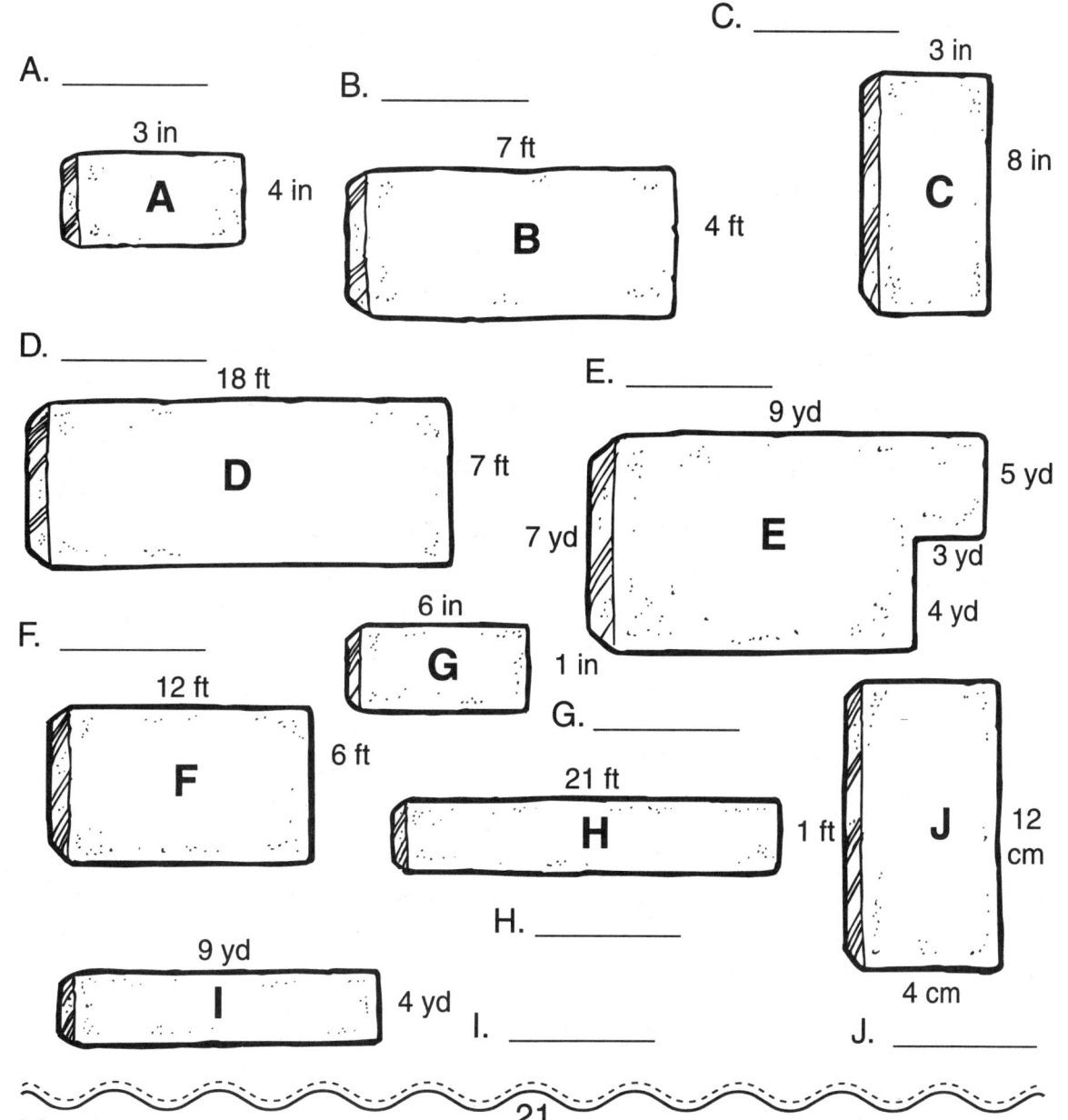

A. _____
3 in
A
4 in

B. _____
7 ft
B
4 ft

C. _____
3 in
C
8 in

D. _____
18 ft
D
7 ft

E. _____
9 yd
5 yd
7 yd
E
3 yd
4 yd

F. _____
12 ft
F
6 ft

6 in
G
1 in

G. _____
21 ft
H
H. _____

J
1 ft
12 cm
4 cm

9 yd
I
4 yd
I. _____

J. _____

© Grace Publications 21 GP-75063 Learning Math With Bible Heroes

Liquid measurement

Jars of Oil

A woman came to Elisha because she was worried about not having enough money to pay her bills. He asked her what she had at home. She said she had a little oil. Elisha told her to gather empty jars from her neighbors and fill them with oil. As she poured oil into the jars, the oil did not run out until she had filled all the jars. Help fill the jars below with oil.

1 pint = 2 cups   1 quart = 2 pints   1 gallon = 4 quarts

How many cups of oil does she need for each of these jars?

A. _____   B. _____   C. _____

D. _____   E. _____   F. _____

G. _____   H. _____   I. _____

© Grace Publications   22   GP-75063 Learning Math With Bible Heroes

Weight

Samson's Weighty Problems

Samson was chosen by God to save his people. God told him to never cut his hair. As long as he obeyed God, he was strong.

Samson knows that nine pennies weigh about one ounce. He also knows that a loaf of bread weighs about one pound. Use this information to help you solve the problems below.

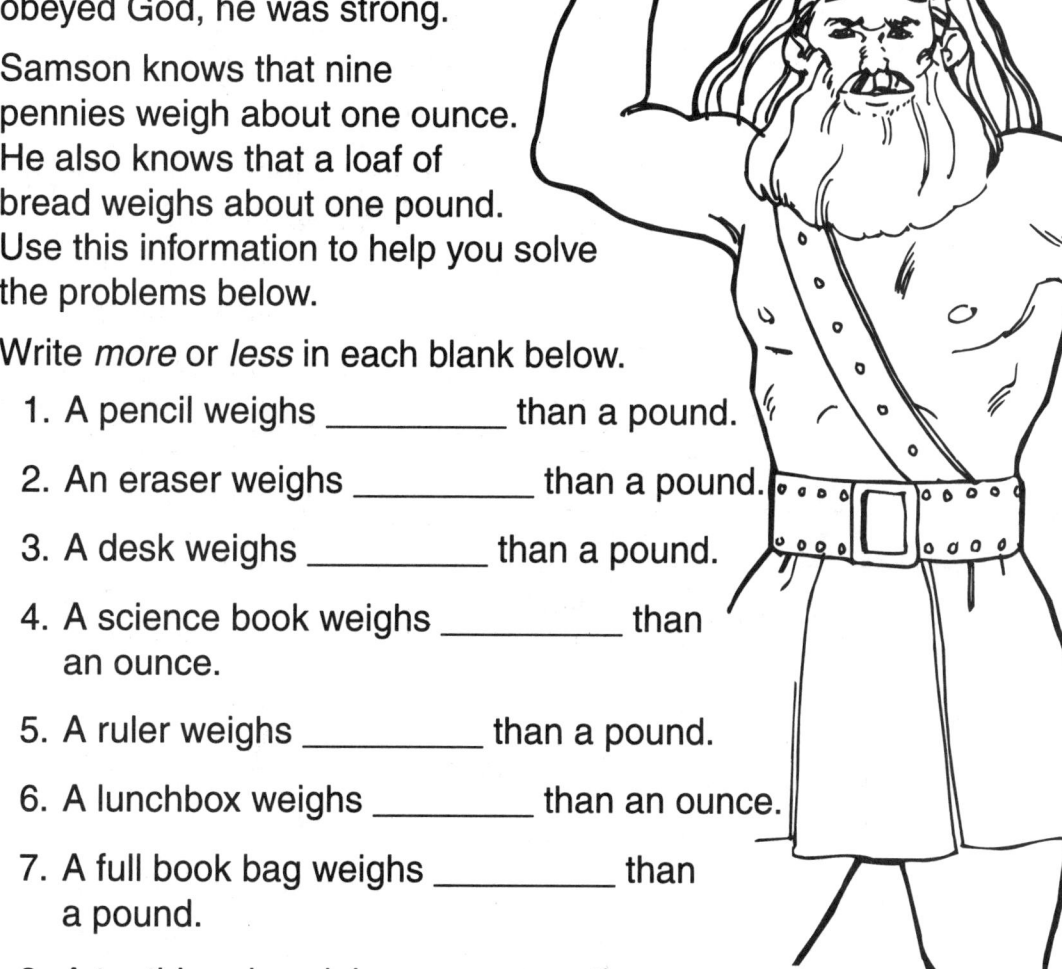

Write *more* or *less* in each blank below.

1. A pencil weighs _____ than a pound.

2. An eraser weighs _____ than a pound.

3. A desk weighs _____ than a pound.

4. A science book weighs _____ than an ounce.

5. A ruler weighs _____ than a pound.

6. A lunchbox weighs _____ than an ounce.

7. A full book bag weighs _____ than a pound.

8. A toothbrush weighs _____ than a pound.

9. Sneakers weigh _____ than an ounce.

10. A marker weighs _____ than a pound.

Geometry

King Solomon

King Solomon chose workers to build a beautiful temple for God. Many basic shapes were used in its construction. Look at the design below. Make a bar graph to show how many of each shape you can find.

Triangles   Squares   Rectangles

Geometry

Solomon's Temple

King Solomon built a beautiful temple so the people could worship God. Many basic shapes were used to build the temple. Help Solomon count all the triangles in the picture below.

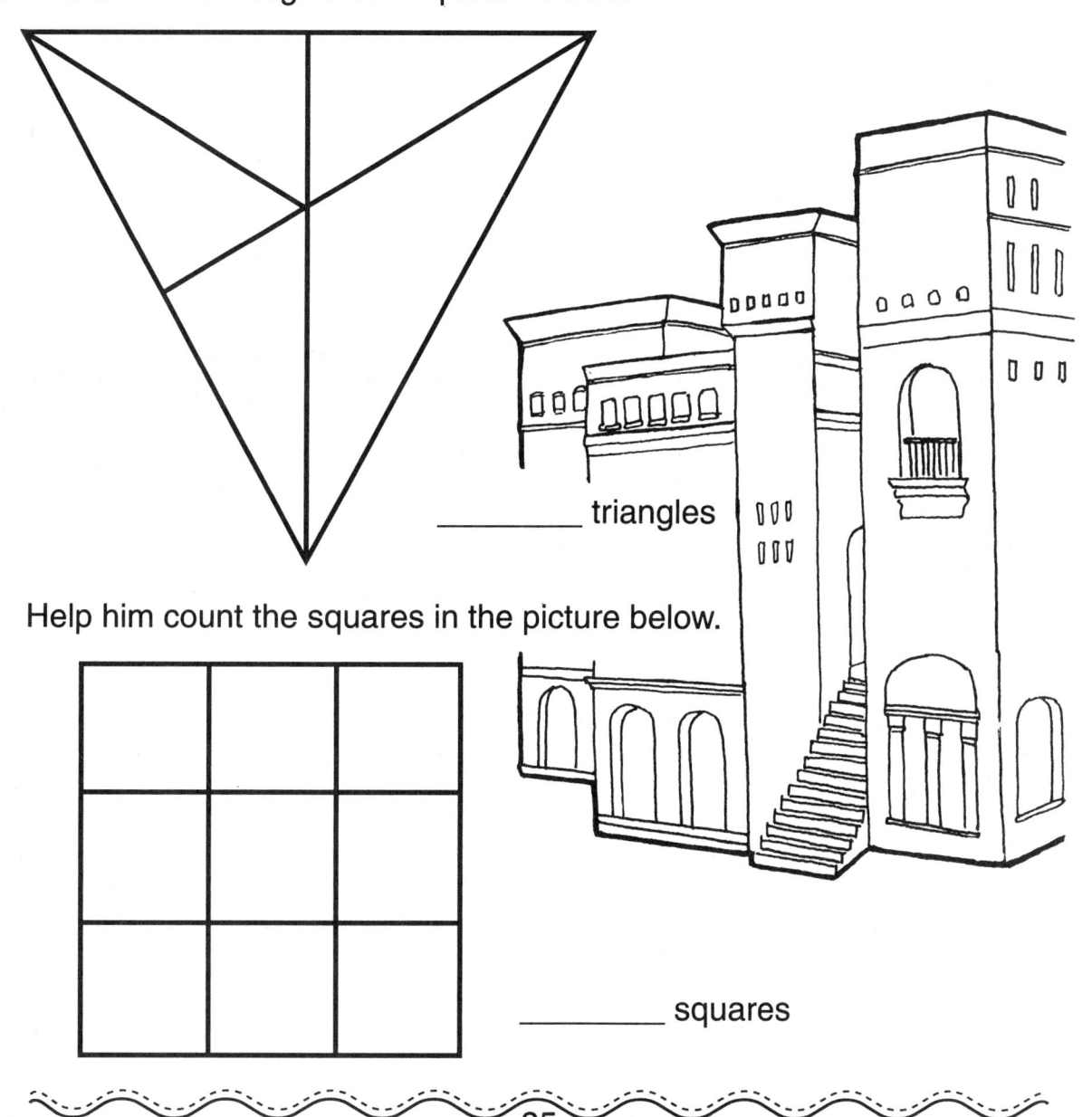

_____ triangles

Help him count the squares in the picture below.

_____ squares

Coordinate geometry

Where Are the Animals?

Noah had many animals to care for on the ark. Some of them liked to wander off. Help Noah locate the wandering animals. Write the coordinates where the animals can be found.

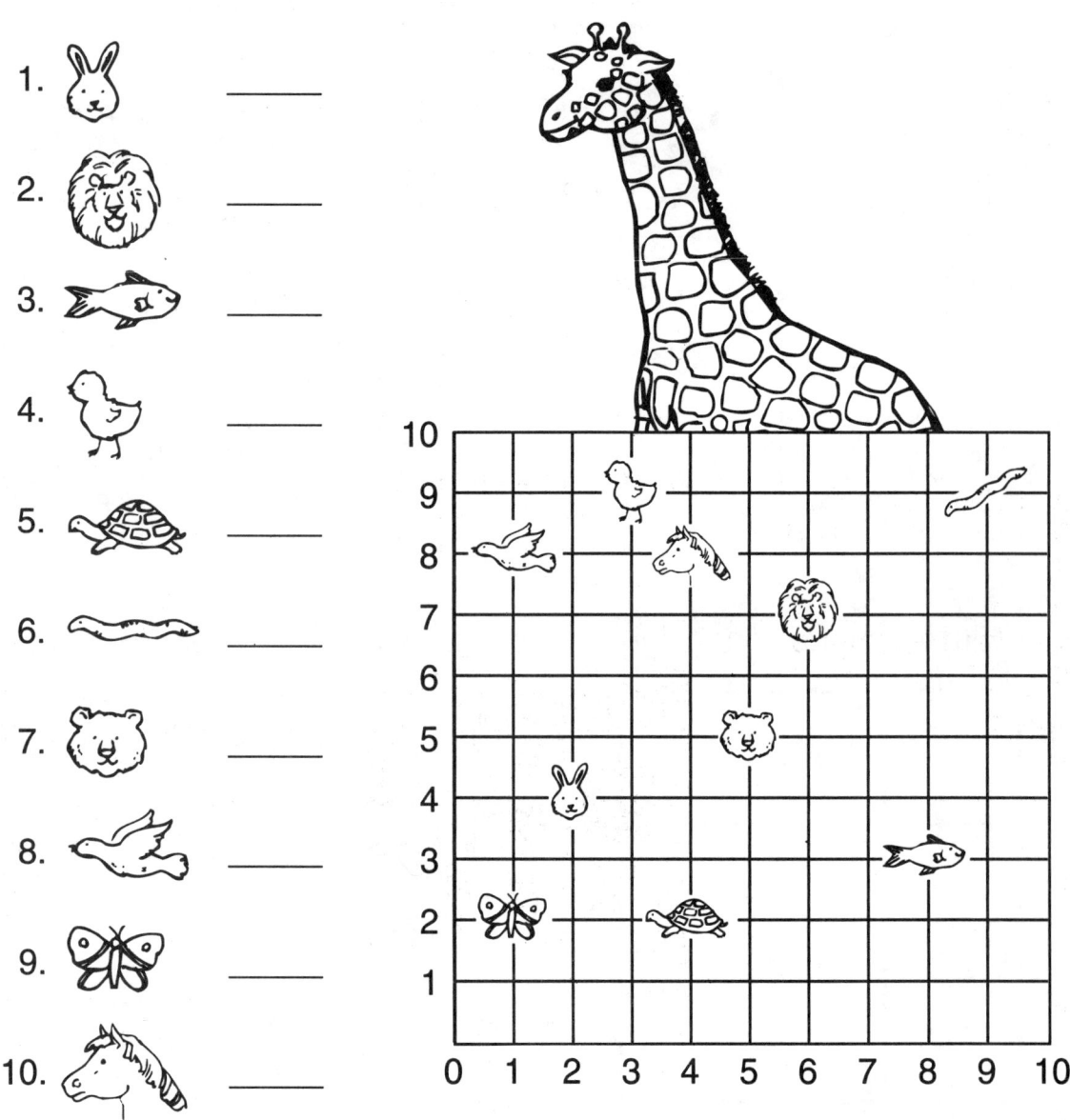

1. _____
2. _____
3. _____
4. _____
5. _____
6. _____
7. _____
8. _____
9. _____
10. _____

Pull-Out Answers

Page 2
1. 6 + 5 = 11
2. 6 + 5 + 7 = 18
3. 5 + 11 = 16
4. 6 − 4 = 2
5. 7 − 6 = 1
6. 14 − 5 = 9
7. 8 + 4 = 12
8. 6 + 5 = 11
9. 4 + 4 + 3 = 11
10. 7 − 4 = 3
11. 5 + 6 = 11
12. 7 − 5 = 2
13. 9 − 4 = 5
14. 5 − 5 = 0
15. 6 + 6 = 12

Page 3
A = 23, B = 37, C = 36, D = 27, E = 20; D should be colored green.

Page 4
A = 18, B = 16, C = 17, D = 19, E = 12, F = 21, G = 24, H = 15

Page 5

A.
11	4	9
6	8	10
7	12	5

B.
4	3	8
9	5	1
2	7	6

C.
9	14	7
8	10	12
13	6	11

D.
12	5	10
7	9	11
8	13	6

E.
4	9	8
11	7	3
6	5	10

F.
6	11	10
13	9	5
8	7	12

Page 6
Answers will vary.

13:
6	6	1
5	🐟	8
2	7	4

12:
5	5	2
7	🐟	5
2	4	6

11:
3	6	2
7	🥜	4
5	4	2

10:
5	2	3
1	🥄	9
2	4	4

Page 7
A. 15 + 4 = 19 + 8 = 27 − 6 = 21 + 7 = 28
B. 8 + 5 = 13 + 5 = 18 − 3 = 15 + 6 = 21
C. 7 + 3 = 10 + 4 = 14 − 7 = 7 + 8 = 15
D. 11 − 4 = 7 + 6 = 13 − 3 = 10 + 10 = 20
E. 14 − 7 = 7 + 9 = 16 − 7 = 9 + 9 = 18

Page 8

4	7	3	2	2	6	
				9	4	6
7	3	3		4	8	3
6		9	0	0	0	7
0		0		7	8	2	0
8	1	9	6	3	4	
	9		8		3		6	4
	3	2	2	4	0	9

Page 9
A. 478 B. 487
C. 748 D. 784
E. 847 F. 874

Page 10
ABRAHAM

Page 11
COVENANTS

Page 12

1. 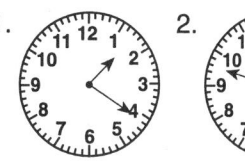 2.

3. 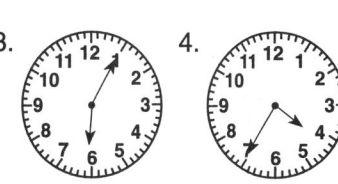 4.

5. 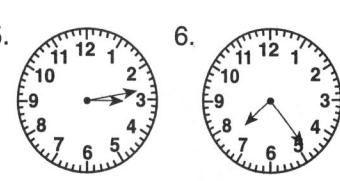 6.

7. 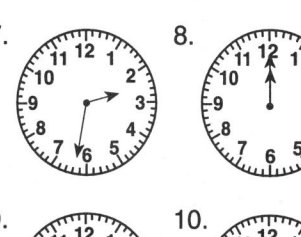 8.

9.  10.

11. 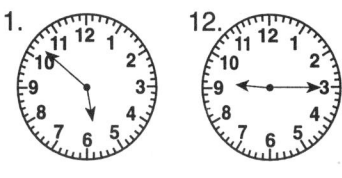 12.

Page 13
1. 3:14, three fourteen
2. 5:47, five forty-seven
3. 8:19, eight nineteen
4. 2:51, two fifty-one
5. 1:14, one fourteen
6. 4:32, four thirty-two
7. 6:30, six thirty
8. 7:26, seven twenty-six
9. 9:42, nine forty-two

Page 14
1. grapes, corn, fish
2. carrots and apples or raisins and corn
3. grapes, seed, carrots, corn
4. raisins, apples
5. grain, apples
6. raisins, fish, seed, grapes

Page 15
1. $5.00, no
2. $2.53, no
3. $1.92, yes
4. $1.44, no
5. $4.31, no
6. $3.19, yes
7. $3.92, no
8. $1.98, no
9. $0.96, yes
10. $1.02, no

Page 16
Answers will vary.

Page 17
1. 954
2. 693
3. 541
4. 424
5. 1019
6. 463
7. 1334
8. 1512
9. 636
10. 1416
11. 869
12. 879
13. 610
14. 1101

Page 18
SLINGSHOT

Page 19
FOREHEAD

Page 20
Answers will vary.

Page 21
A. 14 in
B. 22 ft
C. 22 in
D. 50 ft
E. 34 yd
F. 36 ft
G. 14 in
H. 44 ft
I. 26 yd
J. 32 cm

Page 22
A. 32 c.
B. 6 c.
C. 8 c.
D. 24 c.
E. 8 c.
F. 1 c.
G. 5 c.
H. 12 c.
I. 24 c.

Page 23
1. less
2. less
3. more
4. more
5. less
6. more
7. more
8. less
9. more
10. less

Page 24
17 triangles, 4 squares, 7 rectangles

Page 25
12 triangles, 14 squares

Page 26
1. 2, 4
2. 6, 7
3. 8, 3
4. 3, 9
5. 4, 2
6. 9, 9
7. 5, 5
8. 1, 8
9. 1, 2
10. 4, 8

Page 27
1. 14
2. 8
3. 11
4. 3
5. 6
6. 7
7. 5
8. 2
9. 20
10. 4
BAALS

Page 28
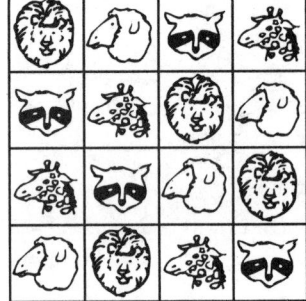

Page 29
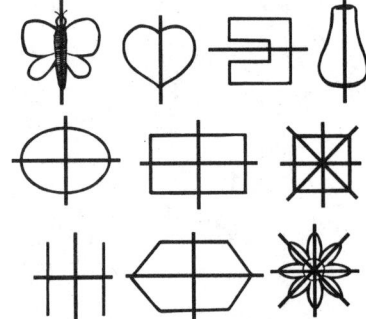

Pages 30-31

1.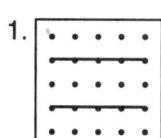
2.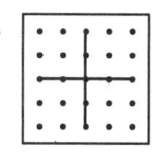
3.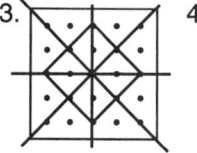
4.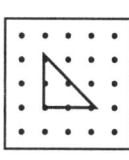
5.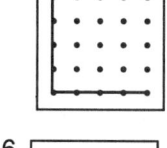
6., 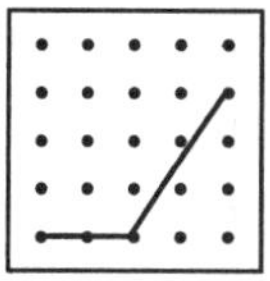 7.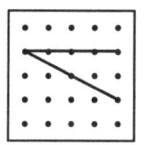

8.-13. Answers will vary.

Page 32
GOLDEN CALF

Page 33
TEN COMMANDMENTS

Page 34
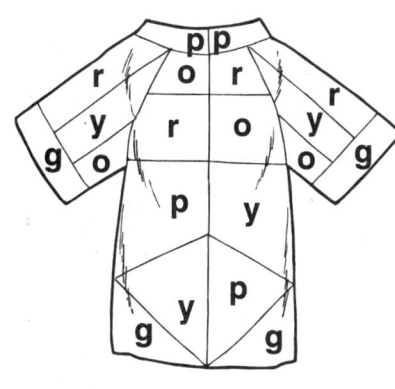

Page 35

1. 9
2. 8
3. 8
4. 4
5. 4
6. 8
7. 2
8. 5
9. 5
10. 8
11. 2
12. 6
13. 6
14. 5
15. 2
16. 9

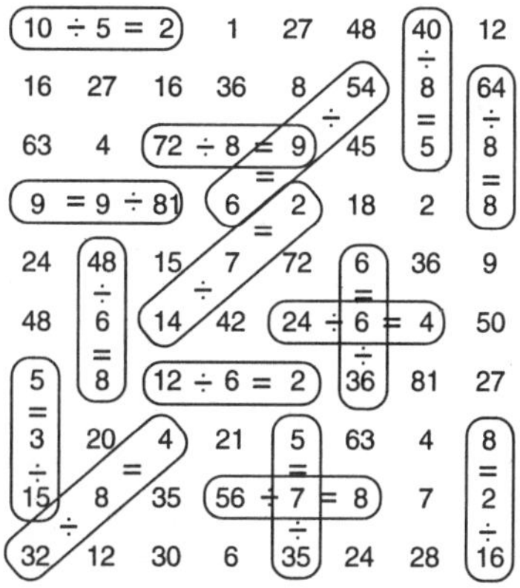

(10 ÷ 5 = 2) 1 27 48 (40 ÷ 8 = 5) 12
16 27 16 36 8 (54 ÷) (64 ÷ 8 = 8)
63 4 (72 ÷ 8 = 9) 45
(9 = 9 ÷ 81) 6 2 18 2
24 (48 ÷ 6 = 8) 15 7 72 (6 = 36) 9
48 14 42 (24 ÷ 6 = 4) 50
(5 = 3 ÷) (12 ÷ 6 = 2) 36 81 27
15) 20 4 21 (5 = 63) 4 (8 = 2 ÷ 16)
32 8 35 (56 ÷ 7 = 8) 7
32 12 30 6 (35) 24 28

Page 36
REBEKAH

Page 37
RAHAB

Page 38
A. 2/6, B. 3/6, C. 1/6

Page 39
A. 5/6 B. 7/8
C. 3/4 D. 8/10
E. 4/5 F. 8/9
G. 3/5 H. 5/7
I. 2/3

Page 40
SHADRACH, MESHACH, and ABEDNEGO

Page 41
A. 1/3 B. 2/7
C. 2/4 D. 4/8
E. 1/6 F. 2/6
G. 5/9 H. 2/10
I. 2/9 J. 5/10
K. 3/5 L. 3/8

Page 42
1. $2\frac{1}{6}$ 2. $2\frac{3}{7}$
3. $6\frac{3}{4}$ 4. $4\frac{4}{5}$
5. 9 6. $1\frac{2}{9}$
7. $1\frac{1}{12}$ 8. $1\frac{1}{3}$
9. $5\frac{1}{4}$ 10. $3\frac{2}{3}$
11. $1\frac{3}{4}$ 12. $3\frac{1}{6}$
13. $3\frac{7}{8}$ 14. $6\frac{1}{2}$

Page 43
PSALMS

Page 44
1. Wednesday
2. Monday
3. 4, 10, 13, 6, 9
4. 5
5. 42
6. line

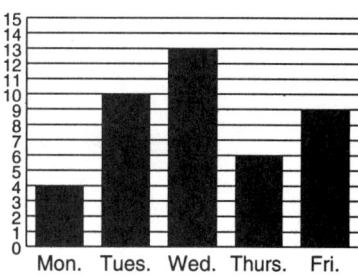

Geometry

The True God

King Ahab built statues of wood, metal, and stone. He wanted all the people to worship these statues. God sent Elijah to the people to show them who the true God was. Answer the questions and use the code to find out the name of King Ahab's statues.

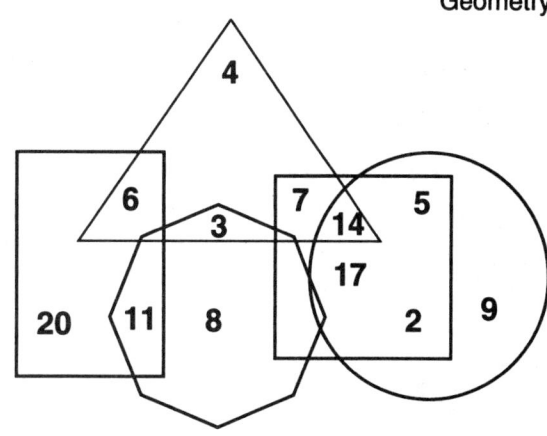

1. Which number is in the circle, triangle, and square? _____ S

2. Which number is only in the octagon? _____ A

3. Which number is in the octagon and the rectangle? _____ I

4. Which number is in the triangle and the octagon? _____ A

5. Which number is in the rectangle and the triangle? _____ B

6. Which number is in the square and the triangle? _____ P

7. Which odd number less than ten is in the circle and the square? _____ L

8. Which even number less than five is in the circle and the square? _____ A

9. Which even number greater than ten is not in the circle, square, or triangle? _____ J

10. Which number is only in the triangle? _____ R

____ ____ ____ ____ ____
 6   3   8   5   14

27

© Grace Publications   GP-75063 Learning Math With Bible Heroes

Problem solving

Noah's Puzzle

Noah had many animals on the ark. When he put them in stalls, he had to be careful. Help Noah put the animals below in the stalls. Cut out the animals and place them in the grid. Only one of each animal may be in each row, column, or diagonal.

Symmetry

God's World Has Symmetry

When God created the world, he created many beautiful shapes in nature. Many of these shapes are symmetrical. Draw lines of symmetry in the shapes below.

Draw one line of symmetry.

Draw two or more lines of symmetry.

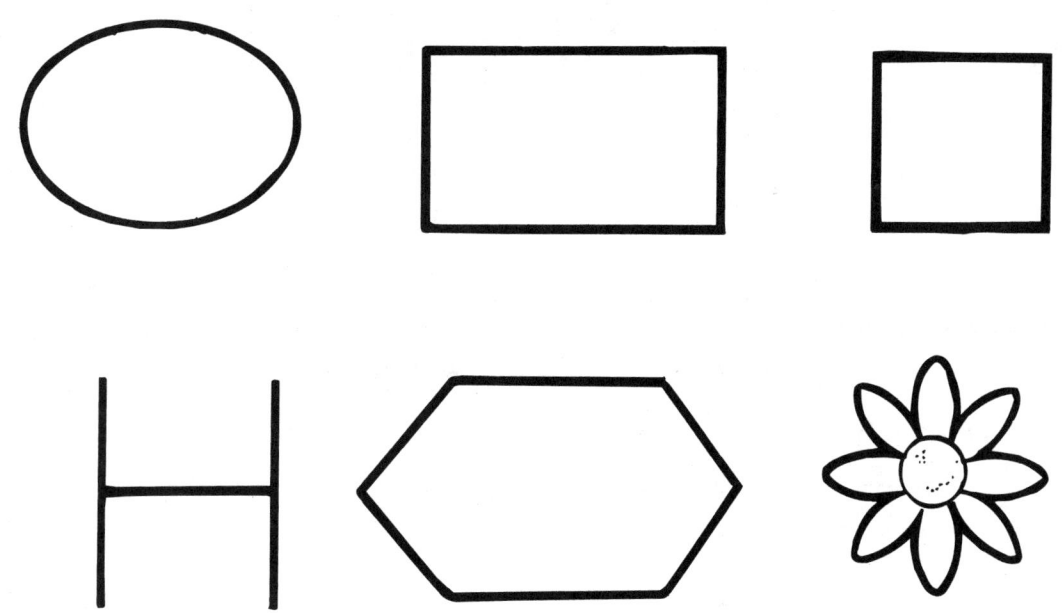

Geometry

Josiah the Boy King

When Josiah was only eight years old, he became king. He saw that God's temple was old and broken so he decided to rebuild it. The Book of the Law of God was found in the temple. King Josiah told the people that they must follow God's laws.

Follow the laws of geometry when you build the shapes on the geoboards below and on page 31.

1. Parallel lines are lines in the same plane that never meet. Draw another line that is parallel.

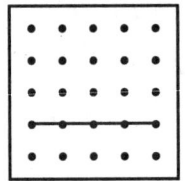

2. Intersecting lines are lines that cross at one point. Draw an intersecting line.

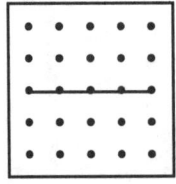

3. Draw two lines of symmetry through this shape.

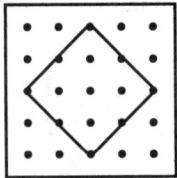

4. An equilateral triangle is a triangle in which all the sides are the same length. Draw an equilateral triangle.

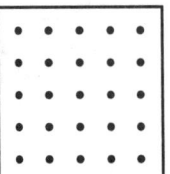

5. Right angles are formed in any corner of a square. Make a right angle.

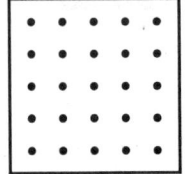

© Grace Publications 30 GP-75063 Learning Math With Bible Heroes

Geometry

Josiah the Boy King continued

6. Make an angle larger than a right angle.

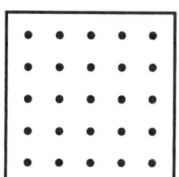

7. Make an angle smaller than a right angle.

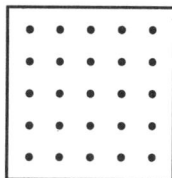

Polygons are closed figures with segments that join at points.

8. Make a polygon with 3 sides.

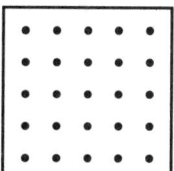

9. Make a polygon with 4 sides.

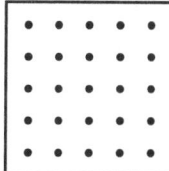

10. Make a polygon with 5 sides.

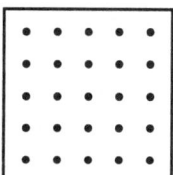

11. Make a polygon with 6 sides.

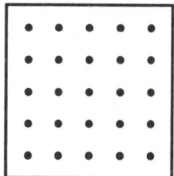

12. Make a polygon with 7 sides.

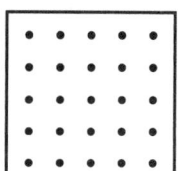

13. Make a polygon with 8 sides.

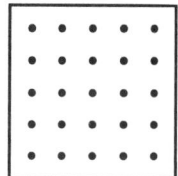

Mount Sinai

Multiplication

While Moses was up on Mount Sinai, the people thought that he was not coming back down. They took all of their gold earrings and melted them down. Solve the problems below. Then use the code to discover what the people made. (Some letters won't be used.)

K. 2 × 8

D. 6 × 7

A. 7 × 5

N. 9 × 9

C. 8 × 7

L. 8 × 3

M. 4 × 4

F. 8 × 9

O. 7 × 7

O. 6 × 8

N. 6 × 6

R. 9 × 3

B. 5 × 2

A. 5 × 9

V. 3 × 4

H. 11 × 7

I. 5 × 5

G. 9 × 7

J. 8 × 5

E. 4 × 8

L. 7 × 4

P. 8 × 8

K. 5 × 3

A. 6 × 3

___ ___ ___ ___ ___ ___   ___ ___ ___ ___
63 49 28 42 32 36   56 18 24 72

Multiplication

God's Law

Moses went to Mount Sinai to talk to God. While he was up there, God told him the rules he wanted to give to his people. To find out what these rules are called, solve the multiplication facts below. Then write the letter next to each problem above its answer at the bottom of the page.

8	1	6
2	7	5
4	3	9

P. ☐ × ☐ =   F. ☐ × ☐ =   N. ☐ × ☐ =

M. ☐ × ☐ =   M. ☐ × ☐ =   R. ☐ × ☐ =

T. ☐ × ☐ =   D. ☐ × ☐ =   O. ☐ × ☐ =

E. ☐   C. ☐   H. ☐   M. ☐   T. ☐   A. ☐
  × ☐   × ☐   × ☐   × ☐   × ☐   × ☐

B. ☐   S. ☐   N. ☐   G. ☐   N. ☐   E. ☐
  × ☐   × ☐   × ☐   × ☐   × ☐   × ☐

___ ___ ___   ___ ___ ___ ___ ___ ___ ___ ___ ___ ___ ___
24   21   40   32   18   28   20   54   15   63   10   6   64   72   48

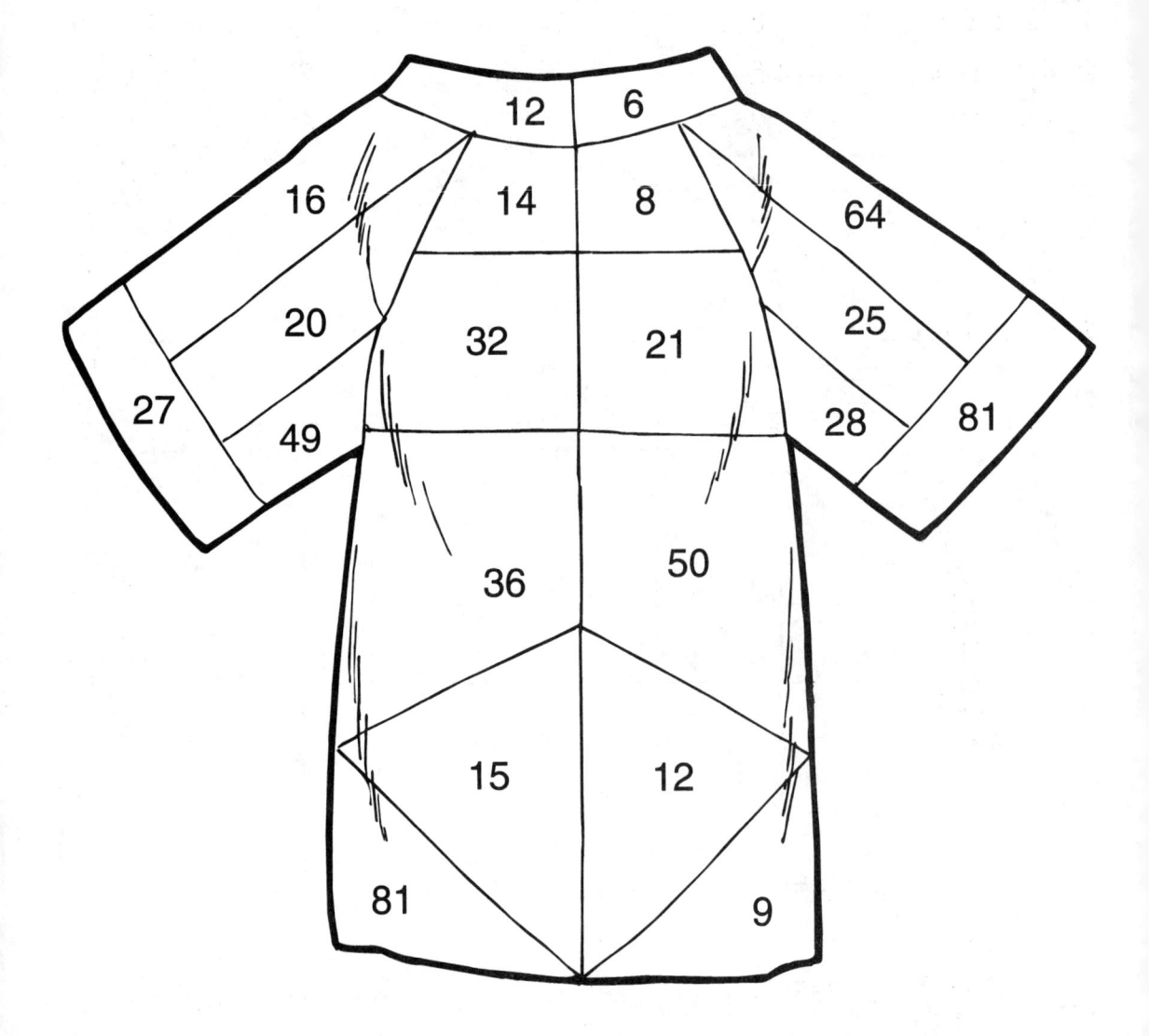

Multiples

Joseph's Coat of "Multiple" Colors

Joseph was Jacob's favorite son. Jacob gave Joseph a brightly colored coat as a special gift. Help decorate Joseph's brightly colored coat using your multiplication skills. Color the multiples as follows:

5s = yellow   6s = purple   7s = orange   8s = red   9s = green

Division

Count on Camels

God's people relied on camels to help them. Camels can go for long periods of time without water. They can also carry heavy loads. Practice your division skills by solving each problem below. Then find each problem on the camel and circle it. Write ÷ and = where they belong.

1. 72 ÷ 8 = 9
2. 48 ÷ 6 = ___
3. 56 ÷ 7 = ___
4. 32 ÷ 8 = ___
5. 24 ÷ 6 = ___
6. 16 ÷ 2 = ___
7. 10 ÷ 5 = ___
8. 15 ÷ 3 = ___
9. 35 ÷ 7 = ___
10. 64 ÷ 8 = ___
11. 14 ÷ 7 = ___
12. 36 ÷ 6 = ___
13. 54 ÷ 9 = ___
14. 40 ÷ 8 = ___
15. 12 ÷ 6 = ___
16. 81 ÷ 9 = ___

10	5	2	1	27	48	40	12
16	27	16	36	8	54	8	64
63	4	72 ÷ 8 = 9			45	5	8
9	9	81	6	2	18	2	8
24	48	15	7	72	6	36	9
48	6	14	42	24	6	4	50
5	8	12	6	2	36	81	27
3	20	4	21	5	63	4	8
15	8	35	56	7	8	7	2
32	12	30	6	35	24	28	16

© Grace Publications

GP-75063 Learning Math With Bible Heroes

Multiplication

A Wife for Isaac

When Abraham and Sarah were very old, God blessed them with a son named Isaac. When Isaac was grown, Abraham sent a servant to find a wife for him. The servant prayed to God to send a girl who could be Isaac's wife to the well. A girl came and offered a drink to the servant and his camel. Solve the problems below. Then use the code to find out the girl's name. (Some letters won't be used.)

O. 62 × 3

R. 87 × 5

E. 97 × 4

E. 43 × 8

A. 23 × 7

E. 35 × 6

J. 94 × 2

K. 52 × 6

L. 19 × 3

B. 49 × 9

S. 38 × 46

A. 27 × 43

H. 69 × 58

P. 47 × 32

T. 85 × 29

___ ___ ___ ___ ___ ___ ___
435 210 441 344 312 1161 4002

Long division

Joshua's Battle of Jericho

After Moses died, God chose Joshua to lead the Israelites to the land that he promised them. The first city they came to was Jericho. It had a wall around it. Joshua sent two men to see what the city was like. The king tried to capture them, but a lady helped hide the men from the king. Solve the problems below. Use the code to find out the name of the woman who helped the men. (Some letters won't be used.)

R. 5)31   B. 3)26   T. 7)22   P. 5)41   M. 2)17

S. 7)39   B. 8)74   J. 7)69   A. 8)28   K. 6)44

A. 4)35   H. 9)57   R. 5)34   H. 4)19   F. 5)29

___ ___ ___ ___ ___
6 R4 8 R3 4 R3 3 R4 9 R2

Fractions

Help Save Daniel

In the story of Daniel and the lions, Daniel disobeyed a law and was thrown into the lions' den. Help Daniel group the lions by coloring their collars. Color 2 collars red. Color 3 collars blue. Color 1 collar green.

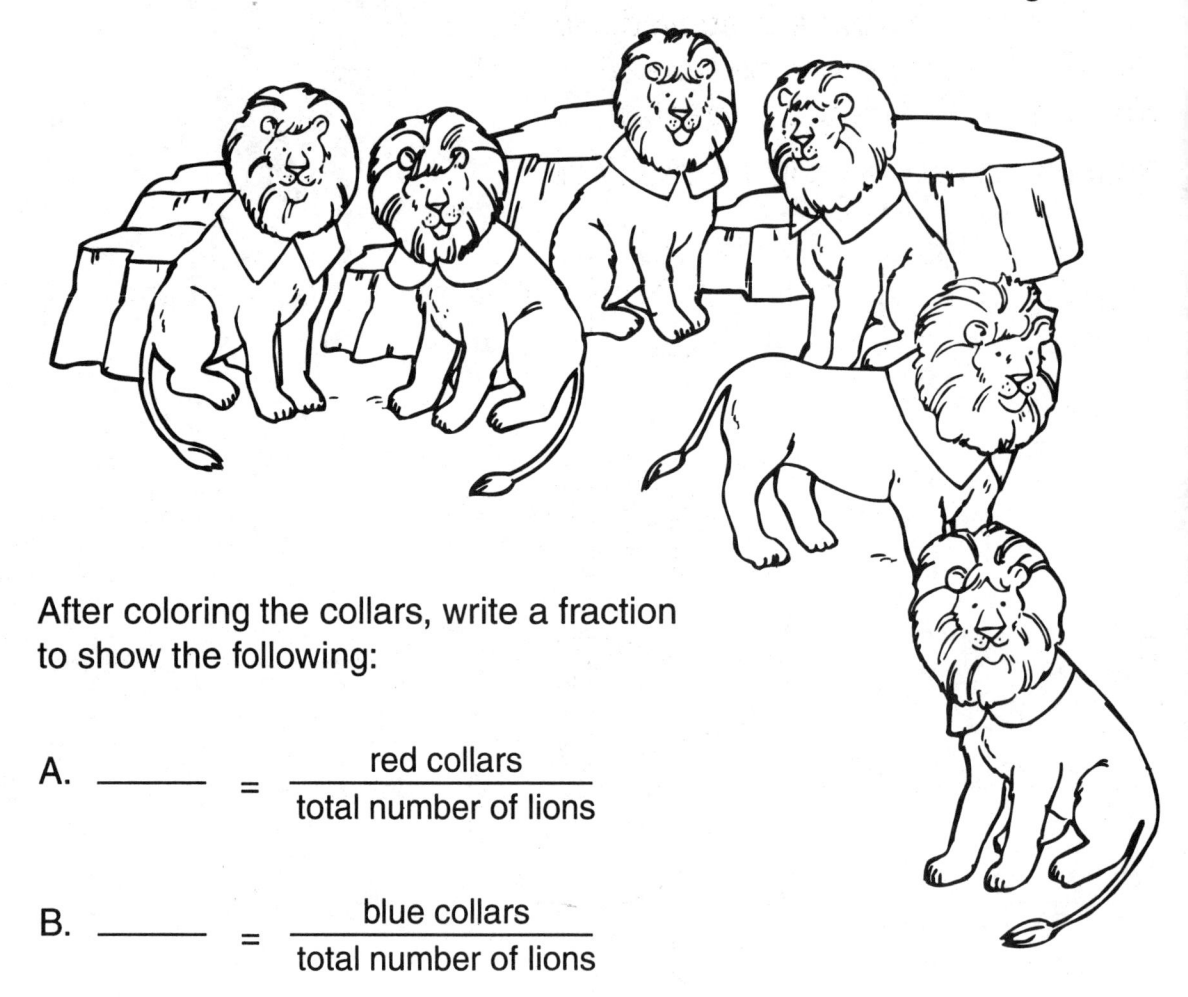

After coloring the collars, write a fraction to show the following:

A. _____ = $\dfrac{\text{red collars}}{\text{total number of lions}}$

B. _____ = $\dfrac{\text{blue collars}}{\text{total number of lions}}$

C. _____ = $\dfrac{\text{green collars}}{\text{total number of lions}}$

© Grace Publications — GP-75063 Learning Math With Bible Heroes

Adding fractions

King Darius Believes

King Darius commanded that every part of his kingdom worship the God of Daniel for he is the living God and his kingdom shall never be destroyed. He rescues and saves, and he works wonders in heaven and on earth.

Shade to show the addition. Then add.

A.

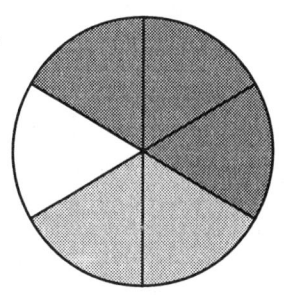

$\frac{2}{6} + \frac{3}{6} = \frac{5}{6}$

B.

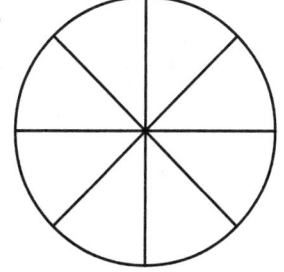

$\frac{2}{8} + \frac{5}{8} =$

C.

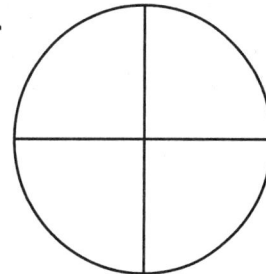

$\frac{1}{4} + \frac{2}{4} =$

D.

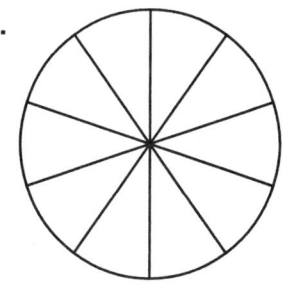

$\frac{6}{10} + \frac{2}{10} =$

E.

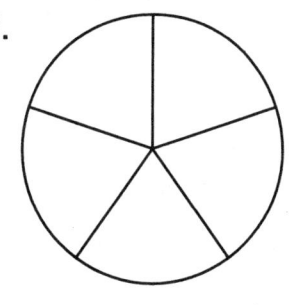

$\frac{3}{5} + \frac{1}{5} =$

F.

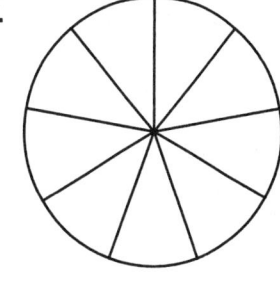

$\frac{5}{9} + \frac{3}{9} =$

Add.

G. $\frac{2}{5} + \frac{1}{5} =$

H. $\frac{3}{7} + \frac{2}{7} =$

I. $\frac{1}{3} + \frac{1}{3} =$

Fiery Furnace

Fractions

The king of Babylon set up a gold statue and told the people to worship it. Three of Daniel's friends refused and vowed they could only worship the true God. The king commanded that they be thrown into a hot fire. To find out who Daniel's friends were, draw a line from each fraction to its equivalent part. Then write the letter above its problem number at the bottom of the page.

1. $\frac{1}{3}$ $\frac{12}{16}$ E

2. $\frac{3}{4}$ $\frac{4}{14}$ G

3. $\frac{5}{7}$ $\frac{20}{28}$ O

4. $\frac{1}{5}$ $\frac{15}{18}$ S

5. $\frac{5}{6}$ $\frac{3}{9}$ A

6. $\frac{2}{8}$ $\frac{5}{20}$ C

7. $\frac{1}{6}$ $\frac{2}{10}$ B

8. $\frac{1}{4}$ $\frac{10}{15}$ N

9. $\frac{3}{5}$ $\frac{3}{18}$ M

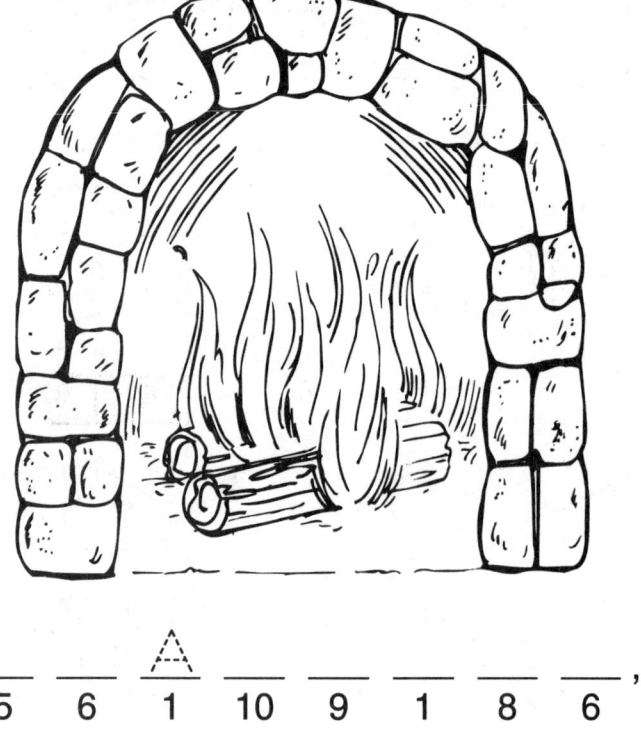

10. $\frac{1}{2}$ $\frac{3}{6}$ D

11. $\frac{2}{7}$ $\frac{9}{15}$ R

12. $\frac{2}{3}$ $\frac{16}{64}$ H

__ __ __ __ __ __ __ __ ,
5   6   1   10   9   1   8   6

__ __ __ __ __ __ __ ,
7   2   5   6   1   8   6

and __ __ __ __ __ __ __ __
      1   4   2   10   12   2   11   3

Subtracting fractions

An Angel of God

When the king of Babylon looked into the furnace to see Daniel's three friends in the fire, he saw four men not three. The king said, "Praise to the God who has sent his angel and saved his servants who trusted in him!" The king said that anyone who says anything against this God would be cut into pieces.

Cross out to show the subtraction. Then subtract.

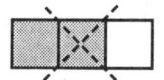

A. $\frac{2}{3} - \frac{1}{3} = \frac{1}{3}$

B. $\frac{5}{7} - \frac{3}{7} =$

C. $\frac{3}{4} - \frac{1}{4} =$

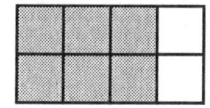

D. $\frac{6}{8} - \frac{2}{8} =$

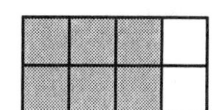

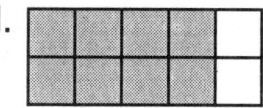

E. $\frac{5}{6} - \frac{4}{6} =$

F. $\frac{4}{6} - \frac{2}{6} =$

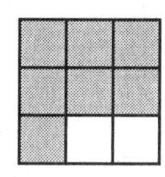

G. $\frac{7}{9} - \frac{2}{9} =$

H. $\frac{8}{10} - \frac{6}{10} =$

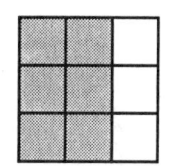

I. $\frac{6}{9} - \frac{4}{9} =$

Subtract.

J. $\frac{9}{10} - \frac{4}{10} =$

K. $\frac{4}{5} - \frac{1}{5} =$

L. $\frac{6}{8} - \frac{3}{8} =$

© Grace Publications   GP-75063 Learning Math With Bible Heroes

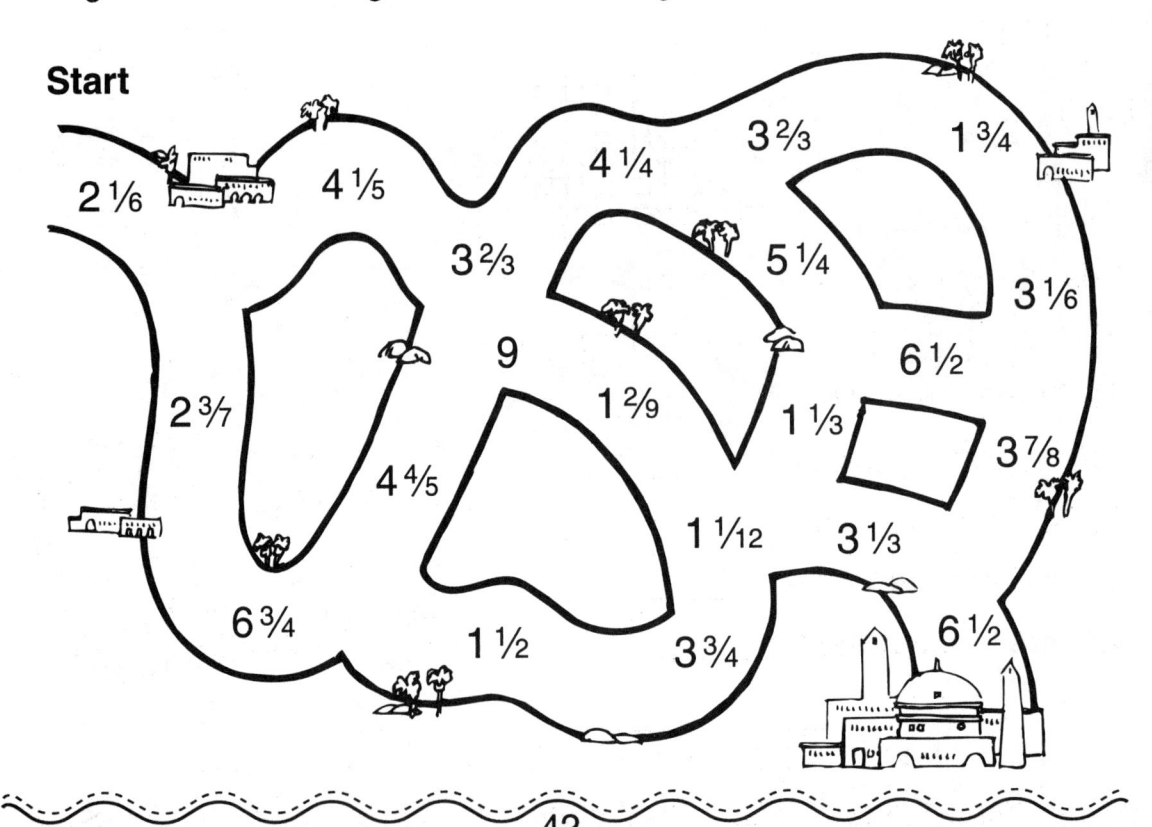

Mixed numbers

Ruth's Journey

After Naomi's husband and sons died, she and her daughter-in-law Ruth decided to move back to Judah. Help Ruth and Naomi find their way by changing the improper fractions to mixed numbers. Begin at Start and color the path that follows the order of the problems.

1. $\frac{13}{6}$ = _____
2. $\frac{17}{7}$ = _____
3. $\frac{27}{4}$ = _____
4. $\frac{24}{5}$ = _____
5. $\frac{18}{2}$ = _____
6. $\frac{11}{9}$ = _____
7. $\frac{13}{12}$ = _____
8. $\frac{4}{3}$ = _____
9. $\frac{21}{4}$ = _____
10. $\frac{11}{3}$ = _____
11. $\frac{7}{4}$ = _____
12. $\frac{19}{6}$ = _____
13. $\frac{31}{8}$ = _____
14. $\frac{13}{2}$ = _____

Start

2 1/6, 4 1/5, 4 1/4, 3 2/3, 1 3/4, 3 2/3, 5 1/4, 3 1/6, 2 3/7, 9, 1 2/9, 6 1/2, 1 1/3, 3 7/8, 4 4/5, 1 1/12, 3 1/3, 6 3/4, 1 1/2, 3 3/4, 6 1/2

© Grace Publications — 42 — GP-75063 Learning Math With Bible Heroes

Decimals

Praise to God

The Israelites sang songs when they worshiped the Lord. These songs helped celebrate God and his goodness. One hundred fifty of these poems that praise God are found in the longest book of the Bible. Answer the problems below. Then use the code to find out what these songs of praise are called. (Some letters will not be used.)

Write a decimal for the colored part of each rectangle.

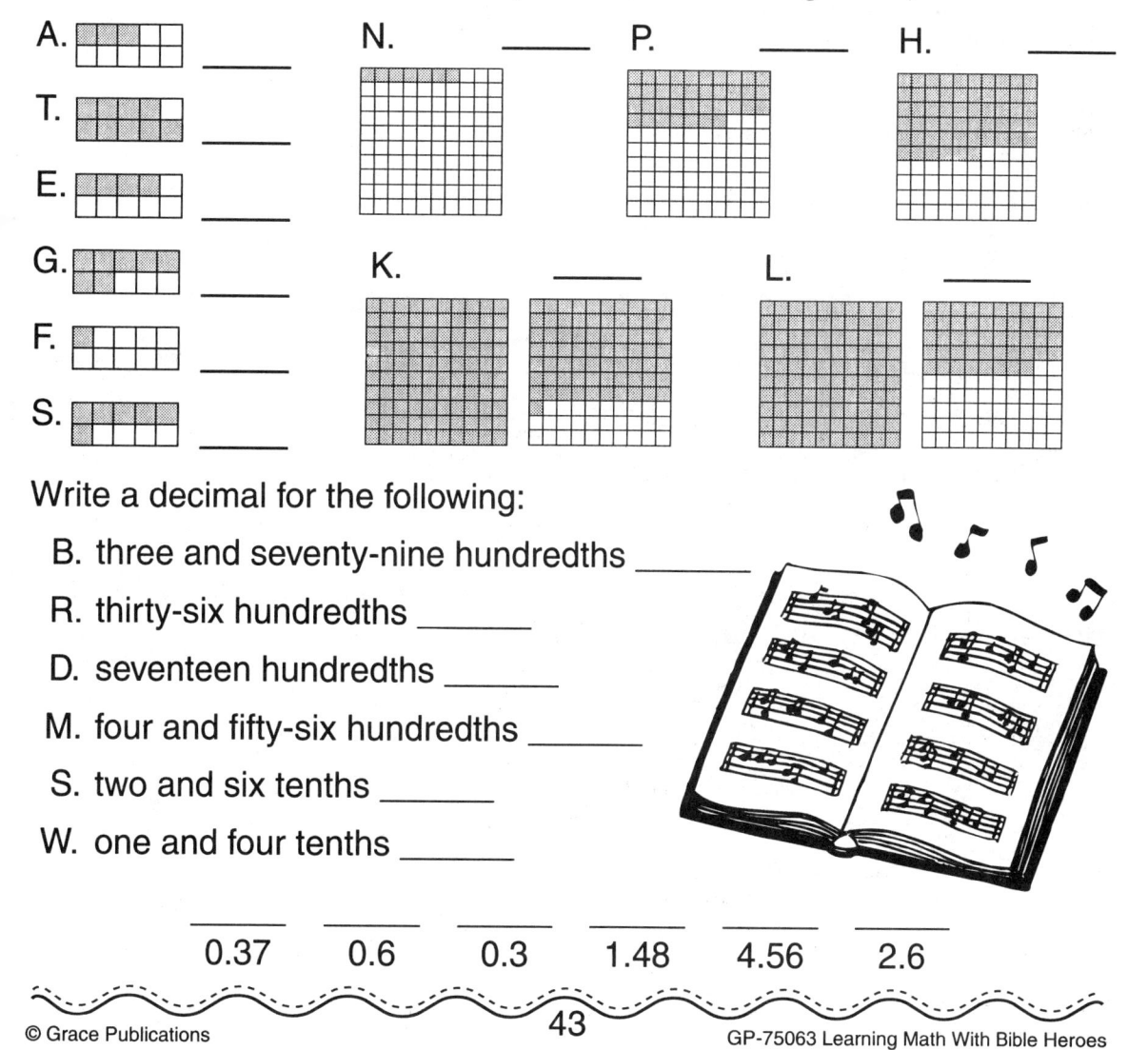

A. _____ N. _____ P. _____ H. _____

T. _____

E. _____

G. _____ K. _____ L. _____

F. _____

S. _____

Write a decimal for the following:

B. three and seventy-nine hundredths _____

R. thirty-six hundredths _____

D. seventeen hundredths _____

M. four and fifty-six hundredths _____

S. two and six tenths _____

W. one and four tenths _____

___ ___ ___ ___ ___ ___
0.37 0.6 0.3 1.48 4.56 2.6

43

Graphs

Miles With Moses

Moses led God's people through the wilderness on their way to the Promised Land. To the right is a graph showing how far they may have traveled. Show your skills by filling in the blanks with the correct information.

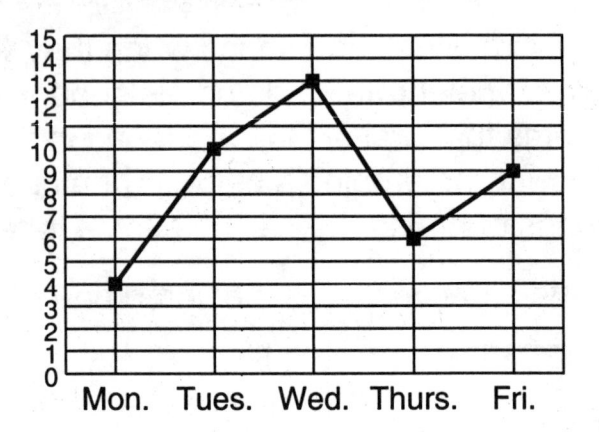

1. On which day did Moses travel the most miles? _____
2. On which day did Moses travel the least amount of miles? _____
3. How many miles did he travel on each day?
   Mon._____ Tues._____ Wed._____ Thurs._____ Fri._____
4. How many days were graphed? _____
5. How many miles were traveled in all? _____
6. This is a _____ graph.

Using the information from the line graph, make a bar graph below. Be sure to label all the parts.

Self-Evaluation

My favorite area of math is _____.

The area of math I am best at is _____.

I did a good job on _____.

I would like to do more _____.

An area I need to improve in is _____.

My plan of action is to _____
in order to make that improvement.

I'm especially interested in _____.

I promise to practice _____.

Signed _____

Date _____

Multiplication Tables

Math tools

2
2 x 1 = 2
2 x 2 = 4
2 x 3 = 6
2 x 4 = 8
2 x 5 = 10
2 x 6 = 12
2 x 7 = 14
2 x 8 = 16
2 x 9 = 18
2 x 10 = 20
2 x 11 = 22
2 x 12 = 24

3
3 x 1 = 3
3 x 2 = 6
3 x 3 = 9
3 x 4 = 12
3 x 5 = 15
3 x 6 = 18
3 x 7 = 21
3 x 8 = 24
3 x 9 = 27
3 x 10 = 30
3 x 11 = 33
3 x 12 = 36

4
4 x 1 = 4
4 x 2 = 8
4 x 3 = 12
4 x 4 = 16
4 x 5 = 20
4 x 6 = 24
4 x 7 = 28
4 x 8 = 32
4 x 9 = 36
4 x 10 = 40
4 x 11 = 44
4 x 12 = 48

5
5 x 1 = 5
5 x 2 = 10
5 x 3 = 15
5 x 4 = 20
5 x 5 = 25
5 x 6 = 30
5 x 7 = 35
5 x 8 = 40
5 x 9 = 45
5 x 10 = 50
5 x 11 = 55
5 x 12 = 60

6
6 x 1 = 6
6 x 2 = 12
6 x 3 = 18
6 x 4 = 24
6 x 5 = 30
6 x 6 = 36
6 x 7 = 42
6 x 8 = 48
6 x 9 = 54
6 x 10 = 60
6 x 11 = 66
6 x 12 = 72

7
7 x 1 = 7
7 x 2 = 14
7 x 3 = 21
7 x 4 = 28
7 x 5 = 35
7 x 6 = 42
7 x 7 = 49
7 x 8 = 56
7 x 9 = 63
7 x 10 = 70
7 x 11 = 77
7 x 12 = 84

8
8 x 1 = 8
8 x 2 = 16
8 x 3 = 24
8 x 4 = 32
8 x 5 = 40
8 x 6 = 48
8 x 7 = 56
8 x 8 = 64
8 x 9 = 72
8 x 10 = 80
8 x 11 = 88
8 x 12 = 96

9
9 x 1 = 9
9 x 2 = 18
9 x 3 = 27
9 x 4 = 36
9 x 5 = 45
9 x 6 = 54
9 x 7 = 63
9 x 8 = 72
9 x 9 = 81
9 x 10 = 90
9 x 11 = 99
9 x 12 = 108

10
10 x 1 = 10
10 x 2 = 20
10 x 3 = 30
10 x 4 = 40
10 x 5 = 50
10 x 6 = 60
10 x 7 = 70
10 x 8 = 80
10 x 9 = 90
10 x 10 = 100
10 x 11 = 110
10 x 12 = 120

11
11 x 1 = 11
11 x 2 = 22
11 x 3 = 33
11 x 4 = 44
11 x 5 = 55
11 x 6 = 66
11 x 7 = 77
11 x 8 = 88
11 x 9 = 99
11 x 10 = 110
11 x 11 = 121
11 x 12 = 132

12
12 x 1 = 12
12 x 2 = 24
12 x 3 = 36
12 x 4 = 48
12 x 5 = 60
12 x 6 = 72
12 x 7 = 84
12 x 8 = 96
12 x 9 = 108
12 x 10 = 120
12 x 11 = 132
12 x 12 = 144

© Grace Publications

GP-75063 Learning Math With Bible Heroes

Mini Geoboards

Below are some extra mini geoboards for you to experiment with. Make some different shapes. Then find the perimeter and area of your shapes.

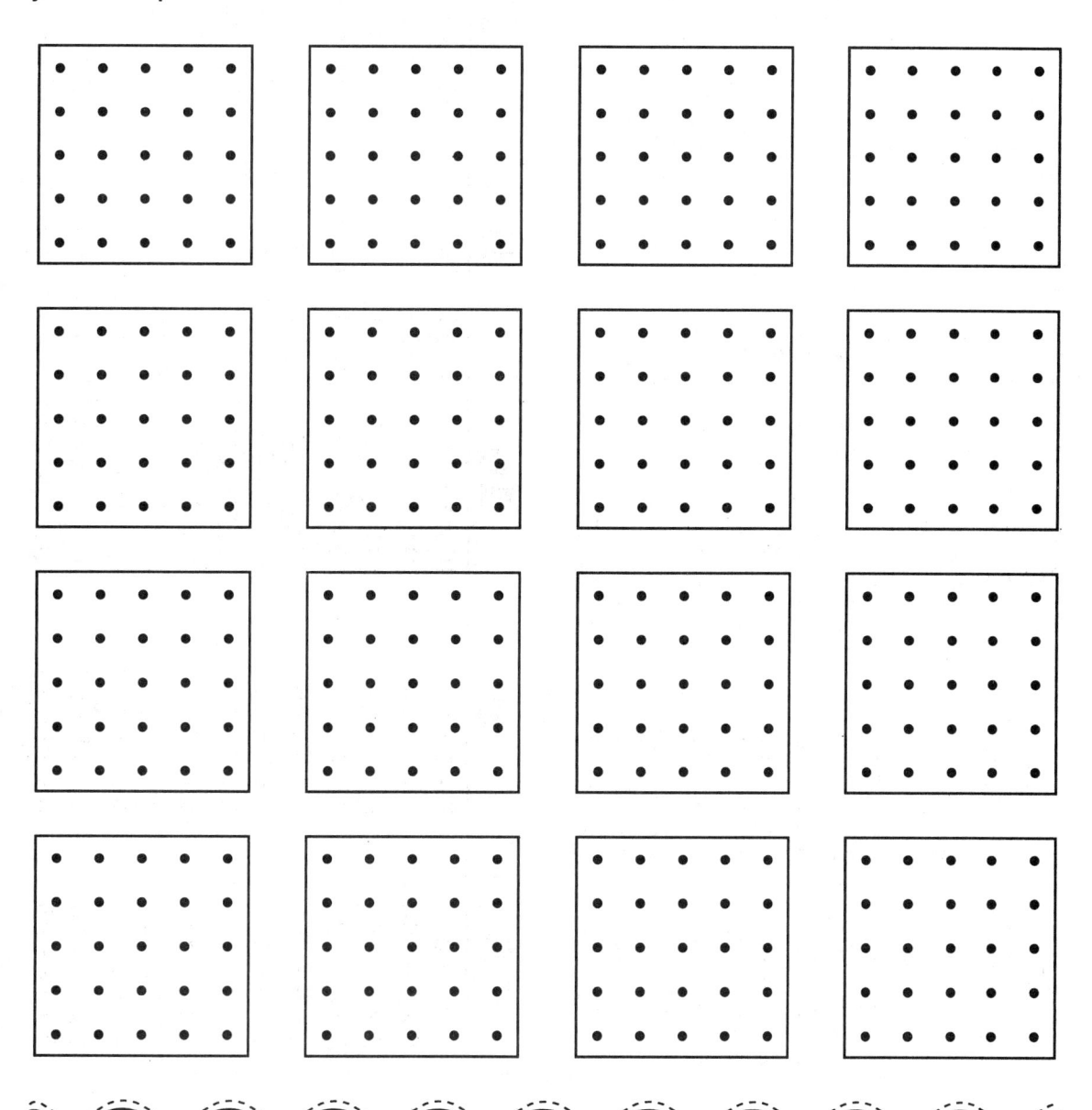

Manipulatives

Mini 100s Charts and Activities

Have children color in multiples of 2, 3, 4, 5, 6, etc. on different charts. Compare the different patterns.

Play Directional Math by giving children a number to start on. Then use arrows to indicate directional moves and the number of moves children should take in that direction.

Example: Start on 23 → 2 ↓ 4 ↖ 1. You should be on 54. Eventually, children should try to follow directions without looking at the chart.

1	2	3	4	5	6	7	8	9	10
11	12	13	14	15	16	17	18	19	20
21	22	23	24	25	26	27	28	29	30
31	32	33	34	35	36	37	38	39	40
41	42	43	44	45	46	47	48	49	50
51	52	53	54	55	56	57	58	59	60
61	62	63	64	65	66	67	68	69	70
71	72	73	74	75	76	77	78	79	80
81	82	83	84	85	86	87	88	89	90
91	92	93	94	95	96	97	98	99	100

1	2	3	4	5	6	7	8	9	10
11	12	13	14	15	16	17	18	19	20
21	22	23	24	25	26	27	28	29	30
31	32	33	34	35	36	37	38	39	40
41	42	43	44	45	46	47	48	49	50
51	52	53	54	55	56	57	58	59	60
61	62	63	64	65	66	67	68	69	70
71	72	73	74	75	76	77	78	79	80
81	82	83	84	85	86	87	88	89	90
91	92	93	94	95	96	97	98	99	100

1	2	3	4	5	6	7	8	9	10
11	12	13	14	15	16	17	18	19	20
21	22	23	24	25	26	27	28	29	30
31	32	33	34	35	36	37	38	39	40
41	42	43	44	45	46	47	48	49	50
51	52	53	54	55	56	57	58	59	60
61	62	63	64	65	66	67	68	69	70
71	72	73	74	75	76	77	78	79	80
81	82	83	84	85	86	87	88	89	90
91	92	93	94	95	96	97	98	99	100

1	2	3	4	5	6	7	8	9	10
11	12	13	14	15	16	17	18	19	20
21	22	23	24	25	26	27	28	29	30
31	32	33	34	35	36	37	38	39	40
41	42	43	44	45	46	47	48	49	50
51	52	53	54	55	56	57	58	59	60
61	62	63	64	65	66	67	68	69	70
71	72	73	74	75	76	77	78	79	80
81	82	83	84	85	86	87	88	89	90
91	92	93	94	95	96	97	98	99	100

© Grace Publications GP-75063 Learning Math With Bible Heroes

Manipulatives

Blank Code Grids

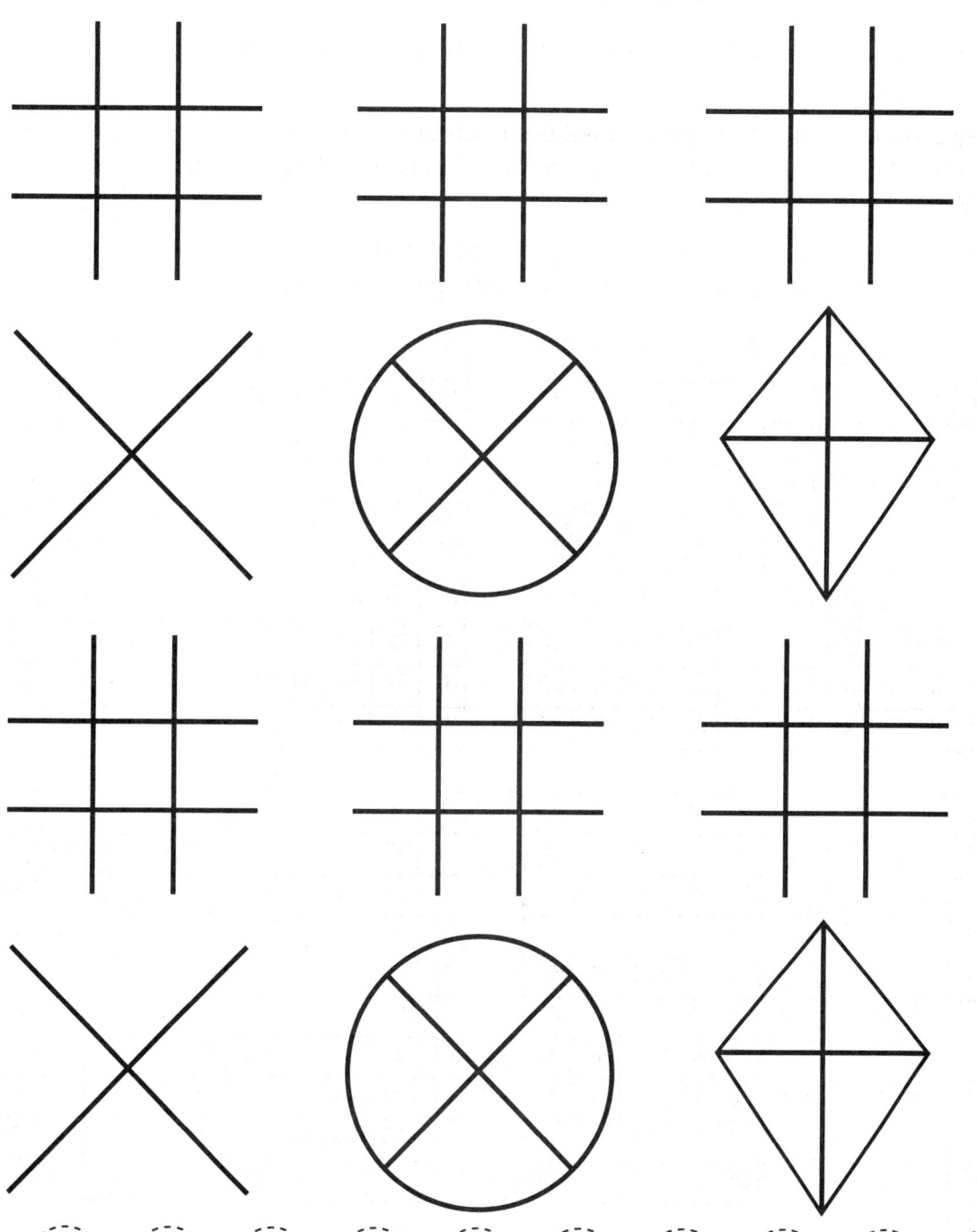

49

© Grace Publications GP-75063 Learning Math With Bible Heroes

Manipulatives

Money Manipulatives

Suggestion: After cutting out your coins, store them in a resealable plastic bag or in an envelope.

© Grace Publications

GP-75063 Learning Math With Bible Heroes

Manipulatives

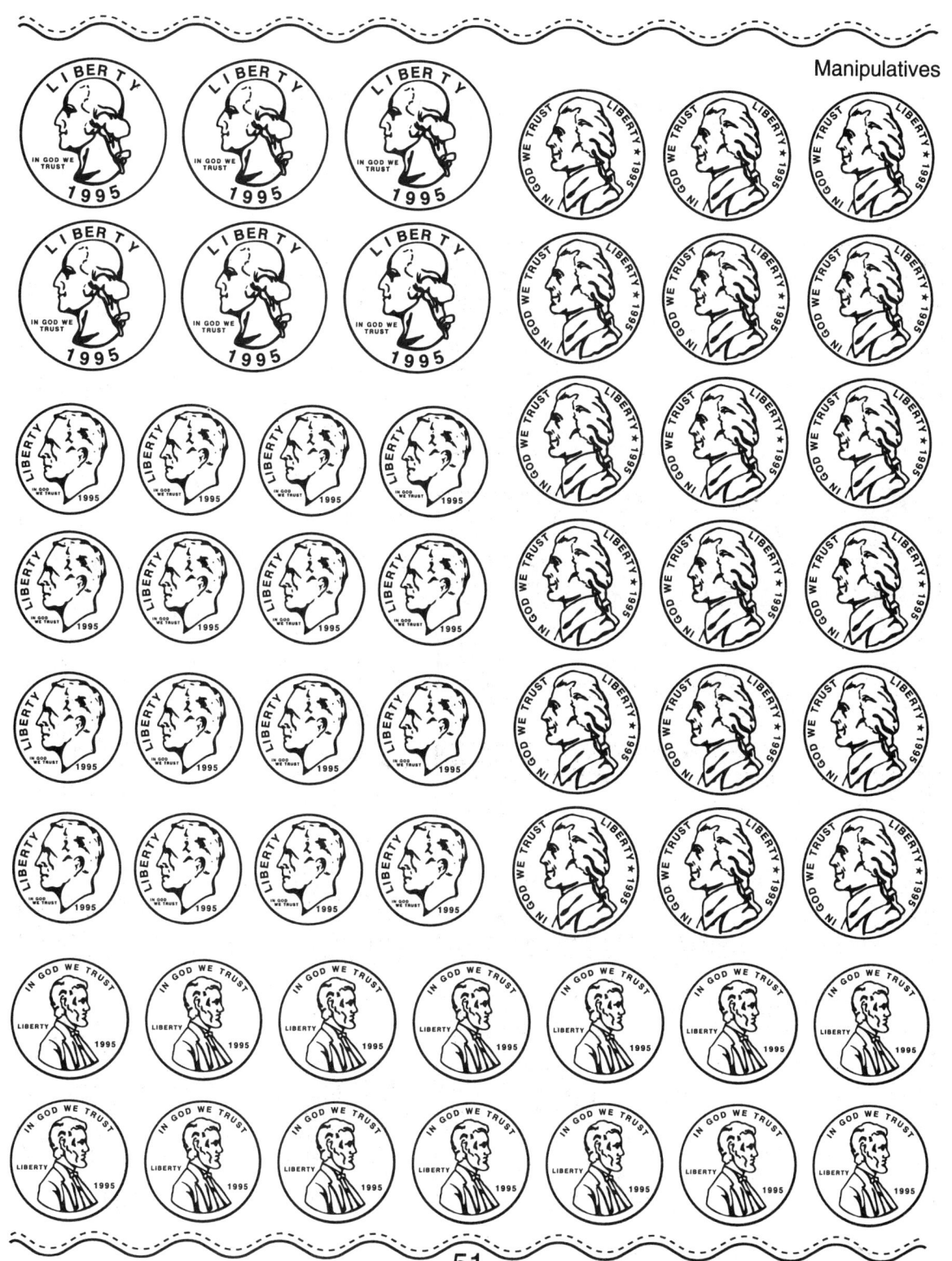

Manipulatives

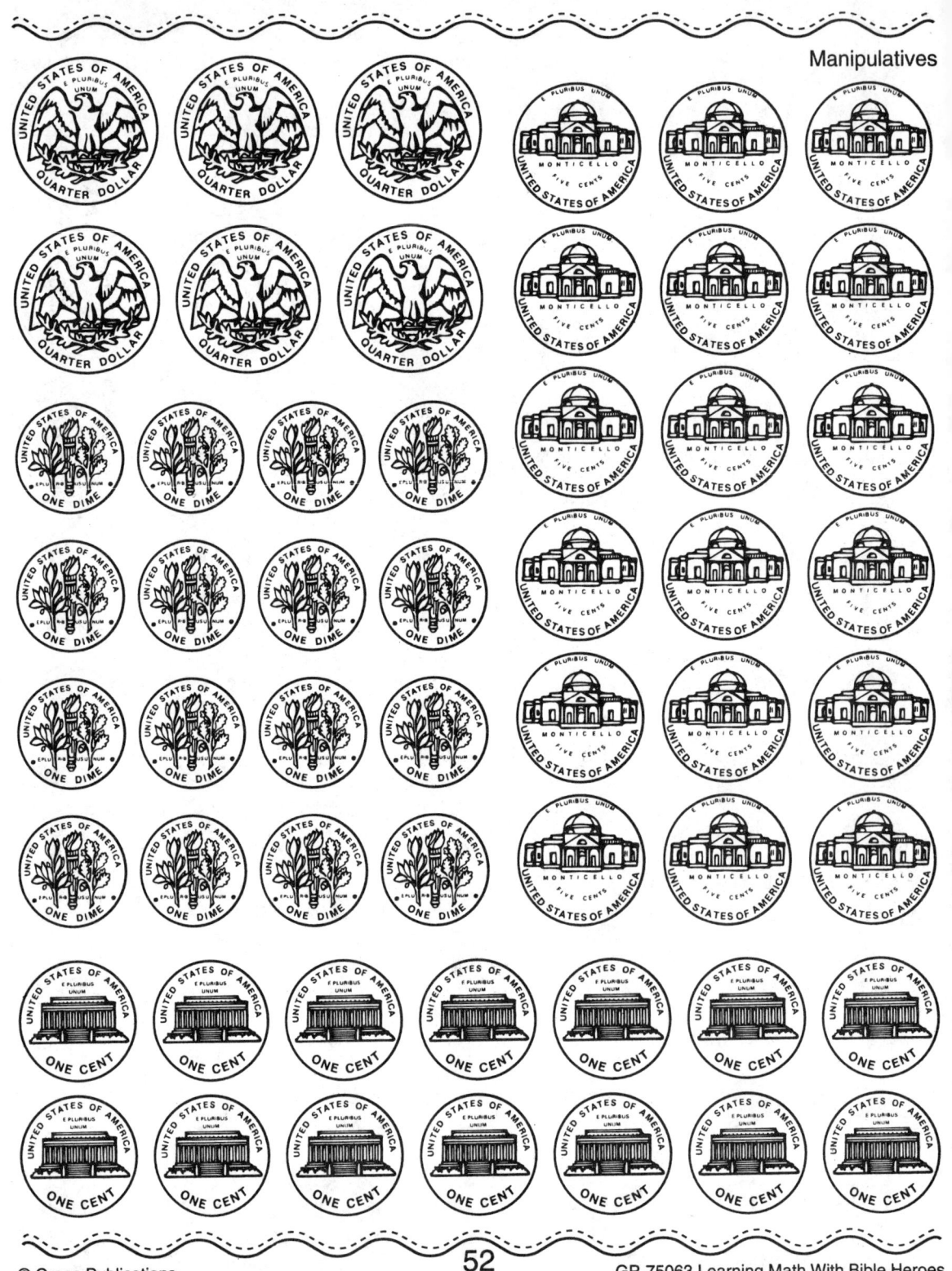